The Measure of Manhattan

THE
City
OF
NEW YORK
AS LAID OUT BY THE COMMISSIONERS
with the
SURROUNDING COUNTRY
By their Secretary and Surveyor
John Randel Jun.r
1814
Scale of Feet 6116 to a Minute
NEW JERSEY
STATE OF NEW JERSEY
HUDSON RIVER
HARLEM RIVER
WEST CHESTER
TOWN OF WEST CHESTER
COUNTY OF WEST CHESTER
EAST RIVER
TOWN OF NEWTOWN
QUEENS COUNTY
HOBOKEN
CITY of JERSEY
HARBOR
BROOKLYN
BUSHWICK
TOWN OF BUSHWICK
BEDFORD
TOWN OF BROOKLYN
KINGS COUNTY
BAY OF NEW YORK
TOWN OF NEW UTRECHT
REMARKS
REFERENCES

The Measure of *Manhattan*

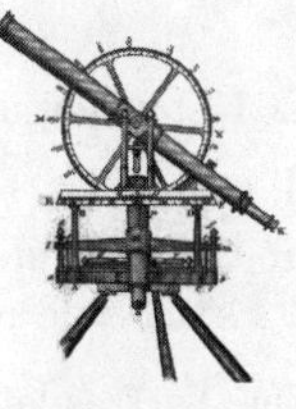

THE TUMULTUOUS CAREER AND SURPRISING LEGACY OF

John Randel Jr., CARTOGRAPHER, SURVEYOR, INVENTOR

Marguerite Holloway

W. W. NORTON & COMPANY
New York | London

Printed in the United States of America
First Edition

Frontispiece: "The City of New York as laid out by the Commissioners, with the Surrounding Country." John Randel Jr.'s 1814 map showed the grid plan that he unveiled in 1811 in its regional context. Courtesy of the New-York Historical Society.

Manufacturing by RR Donnelley, Harrisonburg
Book design by Helene Berinsky
Production manager: Devon Zahn

ISBN 978-0-393-07125-2

W. W. Norton & Company, Inc.
500 Fifth Avenue, New York, N.Y. 10110
www.wwnorton.com

W. W. Norton & Company Ltd.
Castle House, 75/76 Wells Street, London W1T 3QT

1 2 3 4 5 6 7 8 9 0

FOR TOM, AUDEN, AND JULIAN

We shall not cease from exploration
And the end of all our exploring
Will be to arrive where we started
And know the place for the first time.

—T. S. ELIOT

CONTENTS

LIST OF ILLUSTRATIONS

The Measure of Manhattan

PROLOGUE

Big Rock

In the lower level of Riverside Park at 115th Street in New York City lies a long broad rock—part of the hillside, really. When I was a child, it was a mountain. In sunlight it stretched warm, seemingly infinite, offering exploration and even danger, because I might get lost and never find my way home.

I remember visiting that rock when I was twelve and feeling my mind and body in both the past and the present: I could see the rock as I had seen it when I was small, and I could see it as I saw it now that I was taller. It stretched and offered secrets, crevices, a landscape for adventure still, but it didn't seem infinite.

Now I visit that rock with my daughter and son. The children, their schoolmates and teachers call it Big Rock. Sometimes when I go to join them at Big Rock, I overshoot and end up too far north, too far south. The rock's range and extent shift in my memory. I can't consistently find it.

Many people have such a relationship to a tree, a house, a park, an orchard, a street, a copse, a creek—a place where we can feel and see several eras at once, a place where environmental imagination permits simultaneity. In rare wonderful moments, these places allow time travel and size shifting. Different mental maps become transparent and overlapping. Strata of time and scale stir. In these houses or woods or atop a Big Rock, we can truly appreciate time, our transience, and the transience of our perspective.

This book was inspired by John Randel Jr., but it began long before I learned about him. It began when I felt Big Rock big and small and myself small and big all in the same instant. Simultaneity is not déjà vu; it is not a rerun or a feeling of something happening over again in exactly the same way. Simultaneity is instead akin to watching different pictures of the same place, taken at different times and from different perspectives, converge and coexist in a palpable, visceral way. Randel's story took root in a fascination with place, memory, and layers of time that awoke on Big Rock in Riverside Park many years ago.

Big Rock in Riverside Park. Illustration by Patricia J. Wynne.

I

In Which Reuben Skye Rose-Redwood and J. R. Lemuel Morrison Set Out to Find the Imagined City

On a hot June late afternoon in 2004, three people approached the southern drive of Central Park with maps, 10 pounds of high-tech gear, and a growing sense of frustration. They had been climbing rocks, poking in bushes, and scraping at dirt for much of the day, slowly making their way down from the northern reaches of the park. J. R. Lemuel Morrison, a New York City surveyor, wore his reflective orange traffic vest and carried over his shoulder a long pole with a white-and-red GPS unit attached. Reuben Skye Rose-Redwood, then a doctoral student in geography at Pennsylvania State University, and CindyAnn Rampersad, a social geographer now married to Rose-Redwood, passed a sheaf of maps back and forth and took turns carrying a metal detector. It was Rampersad's first day out with Morrison and Rose-Redwood, but she had been listening to tales of their forays for two months and was eager to join them. Central Park's southern drive undulates above 59th Street, above a pond and an ice-skating rink. The park's designers, Frederick Law Olmsted and Calvert Vaux, intended the entrance at 59th Street and Fifth Avenue to be the grand one, the "handsomest" approach. The routes from that corner weave north, leading amblers toward the only straight path in the park, the Mall. Passionately antilinear, Olmsted and Vaux made a single exception so visitors could move directly to the core of the park: the Ramble and the Lake.[1]

The southbound trio stopped several times along the Mall, checking GPS coordinates, scrutinizing maps, examining a rock sitting behind a fence and a PLEASE KEEP OUT sign, and rummaging through a patch of periwinkle until they became worried they would attract onlookers or, worse, someone who worked in the park. Just to the west, a film crew had given Sheep Meadow a nineteenth-century cast by releasing ruminants to populate a scene in a romantic comedy. When they reached the southern end of the Mall and the statue of William Shakespeare, Morrison, Rose-Redwood, and Rampersad checked the maps again. They were looking for the relic of an invisible intersection, one city leaders had planned to build in 1811 but that had never been constructed. For Morrison and Rose-Redwood, finding such a relic had become an obsession—an obsession with a little-explored part of New York City's history, an obsession with those rare moments when past, present, and future overlap, and an obsession with a mysterious man whose remarkable precision was proving as invaluable today as it had two centuries earlier.

Morrison held the range pole steady, turned on the GPS unit, and hoped it would pick up the signal of at least five satellites, which would give him his location to within 5 centimeters. Morrison wears his brown hair long, his glasses rectangular. He is fond of paraphrasing Admiral Grace Hopper: "It is easier to ask for forgiveness than for permission." By laying a digitized map of planned but never realized intersections on top of satellite photographs of the park, Morrison could trace the old city atop the new. But every other invisible intersection they had checked that summer had disappointed. Rose-Redwood felt hot, sweaty, and a bit crestfallen: "You spend hours looking and you don't find anything. And once you look at one spot, you know that you could look again at the same place, but unless you start digging, you are not going to find anything more." Rose-Redwood is five foot eight, dark-haired, with bright blue eyes; he laughs easily and cheerfully discusses philosopher Bruno Latour's inscription devices or Michel Foucault's views on discipline and governmentality.

The summer had not brought Morrison and Rose-Redwood success, but it had delivered adventure. The two had climbed to the summit of Marcus Garvey Park, formerly Mt. Morris Park, at 120th Street and Fifth Avenue. That particular invisible intersection took them up a steep, slippery slope and into a campsite. The occupants "were not the Patagonia-wearing-type campers," Morrison remarked, so he and Rose-Redwood had scrambled quickly away. Working at a long wooden table under the soaring ceiling of the New-York Historical Society library, they had struggled to decipher old documents and sketches. And they had observed the transit of Venus, a rare event during which the second planet can be seen as a small dark dot sailing across the bright surface of the sun. Transits occur in pairs, about eight years apart, every century or so. The 1761 transit was observed by Charles Mason and Jeremiah Dixon shortly before they surveyed the Mason-Dixon line. Morrison had brought his theodolite, a surveying instrument with a telescope, into Central Park in the early morning and set it up at Belvedere Castle.

Rose-Redwood and Morrison had been working together and exploring invisible intersections since late spring, when they had first met at the Landmarks Preservation Commission, a city agency charged with protecting significant historical sites. Rose-Redwood had been invited to give a lecture about New York City's 1811 grid plan, the subject of his master's thesis, and about what Manhattan island looked like pre-grid. The grid plan—much adored and much abhorred, although the abhorrers have been perhaps more vocal—established a trellis for New York City to climb as it grew north from what was then North Street, now Houston Street, to 155th Street. In 1806 the city government had determined that New York was expanding too rapidly, without blueprint. Hoping to avoid the passionate, bitter debates that erupted whenever land sales or boundary issues arose, city politicians asked the state legislature to help, although the politicians adroitly recommended three commissioners who might be appointed to devise a plan. In 1807 Albany did appoint those three men to develop a map for growth, which the commissioners delivered four years later. Although earlier plans for New York had proposed variations on a grid, the 1811 plan envisioned grid on unprecedented scale. With relatively few exceptions, most conspicuously Central Park, the kinked off-kilter route of Broadway, and the D.C.-like angle of St. Nicholas Avenue, New York City flowed into that form, like batter flowing into the grooves of a waffle iron.

Since 2001, Rose-Redwood had been trying to calculate how significantly the landscape had changed as developers filled in the grid. Many people have criticized the grid plan for its topographical insensitivity, its callous leveling of hills that could have conferred character and charm. The grid promoters, lamented one early critic, "are men, as has been well observed, who would have cut down the seven hills of Rome, on which are erected her triumphant monuments of beauty and magnificence, and have thrown them into the Tyber or the Pomptine [*sic*] marshes." Was Manhattan truly flattened by the grid, Rose-Redwood wanted to know, or are the ups and downs of New Yorkers today roughly the same as they were in the early nineteenth century? Using data sets available from the early days of the city, Rose-Redwood created a rough map of a rough and hilly island—more extreme in its undulations than it is today.[2]

Lemuel Morrison listened to Rose-Redwood's afternoon talk at the Landmarks Preservation Commission in May 2004, thinking to himself, *This guy is a surveyor and he doesn't even know it!* Afterward, he asked Rose-Redwood whether any of the original survey markers that delineated the grid had survived. Rose-Redwood expected that most of them had been lost or destroyed as the city was constructed and reconstructed. But if any such markers had escaped detection or destruction, he mused, they would probably be in Central Park or perhaps another Manhattan park. He invited anyone who was interested in grid history to ramble with him and seek markers that summer. Morrison e-mailed him a few days later.

Morrison and Rose-Redwood wanted to find a token from imagined New York, from the years when New Yorkers gave form to the dream of a grand metropolis. They wanted to find physical evidence for a plan some scholars have called "the single most important document in New York City's development." For Rose-Redwood, such a marker represented a missing piece of scholarship, of history. Not only was he fascinated by the island's original topography, he had come to view the grid plan in a way that differed somewhat from the views of other scholars, and he wanted to discover something tangible about this project he had spent three years researching. For Morrison, the search for historical markers is intuitive. Surveyors are necessarily rooted in the past; they locate former boundaries or evidence about old boundaries, and they correlate those boundaries with those of the present and the future. "Mostly we try to find what is old, and try to be creative. So we love history. Finding old things, that is what we do," he said. The two men perceived a marker from the 1811 plan, should they discover one, as a kind of pivot—a place where the past, present, and future city were held together.[3]

"I think I am the one who found it," Rose-Redwood later recalled. "But it really doesn't matter." The three seekers had fanned out around the GPS unit, which indicated that they were in the middle of yet another invisible nineteenth-century intersection, one now teeming with bicyclists, joggers, and horse-drawn carriages, one redolent with the prickly smell of manure. Several minutes of searching later, Rose-Redwood gave a shout. A dark brown bolt, about 1 inch square, jutted 3 inches out of a rock. It sat in a bed of lead. "We were really quite euphoric," said Rose-Redwood.

Central Park bolt. Illustration by Patricia J. Wynne.

They examined the bolt, discerning faint marks on its flat top—a kind of surveyor's signature found on most such markers. They debated the bolt's legitimacy. Rose-Redwood felt 85 percent sure it was a grid bolt but wondered if they could get a chemist to test the metal and perhaps date it so they could be positive. "What we need to be very cautious about is that there have been many surveys in Central Park since it was laid out in the 1850s and 1860s," he fretted. "I mean, we might find something that is from a different survey." Morrison was 99 percent sure, because to his mind, the location was too perfectly grid-aligned to be coincidence; they had found other bolts in the park that summer, but none of them corresponded to an 1811 intersection. They also began to feel protective of the bolt. Their concern proved well founded: since their discovery, one corner of the bolt has been chipped. Morrison suspects a parks department's lawn-mowing machine. He and Rose-Redwood continue to worry that further damage will come to the bolt. They have suggested to the parks department that a small clear case be set protectively over it. The bolt would sit inside its box, upon its boulder, amid its city, which would sprawl around it, no longer wild, like poet Wallace Stevens's jar upon its hill.

The man who set the bolt in that rock—or who, more likely, instructed his men to fracture the rock with gunpowder, plug the hole with lead, and anchor a 1-inch-square, 6-inch-long bolt—was named John Randel Jr. Hired in 1808 by the three state commissioners to plan the grid and then in 1810 by the city government to implement the grid, Randel hiked the island's hills, waded through its creeks and marshes, and let the tide rise up to his shoulders for more than a decade as he laid down the grid plan. He measured each block, each street, each avenue with a precision that remains admired and relied on today by engineers, planners, and surveyors such as Morrison.

Despite his important role as inscriber of the grid, Randel, who lived from 1787 to 1865—a long life for that period—has been a historical shadow. A few scholars of cartography, urban planning, and infrastructure have in general briefly described Randel in their books as an eccentric, litigious fellow who was involved with many of the major infrastructure works of his time. Randel was indeed strikingly busy. In addition to surveying Manhattan and pinning the grid to the land, he surveyed and divided portions of wild terrain in upstate New York, and he designed towns there as well. He trudged hundreds of miles, laying out turnpikes and surveying routes for several of the country's earliest railroads: the New Castle & Frenchtown in Delaware and Maryland, the Ithaca & Owego in New York, the Lykens Valley Railroad in Pennsylvania, the Delaware Railroad Co., and the Central of Georgia. For some of the railroads he served as engineer in chief, innovating and improvising solutions for the many problems that arose on these early American ventures. Randel sounded the Hudson River south of Albany, assessing how ships might more easily make the voyage up- and downriver from that important port. One winter he risked his health to map the islands of the St. Lawrence River.

He was among the canal pioneers; he worked on the Delaware & Raritan Canal, the Erie Canal, the Chesapeake & Delaware Canal and did surveys for a Pennsylvania canal. He invented remarkable surveying instruments, an earth-toting tram, and was one of the first to dream up, design, and advocate for an elevated railroad in New York City. He had visions for suburban planning, for traffic flow and roads in Manhattan. He consulted on Baltimore's water-delivery system and New York's sewage system.

Randel also created some of the most beautiful and detailed maps of his time, several of which are cartographic gems. On the nineteenth floor of One Centre Street, in the Manhattan borough president's office, in the small room where the island's official maps are kept and consulted hundreds of times every year, are ninety-two remarkable sheets of paper, carefully stored in four boxes. Each measures 32 by 20 inches; assembled, they would form a map 11 feet wide and 50 feet long, a map depicting the marshes, meadows, inlets, rocks, hills, barns, cider mills, property lines, ice houses, lanes, streets to come, fishermen's shacks, and every other structure on, as well as natural feature of, Manhattan in the early nineteenth century. Another map sits in the collection of the Library of Congress, encased in Mylar. It is a 27-by-39-inch map of Manhattan that is part trompe l'oeil, in which the island sits on a scroll unfolding on top of a map of the northeastern states, while over the boroughs of Queens and Brooklyn a small map of Philadelphia unfurls. Randel artfully crafted three maps as one, and it is the only map known to combine this view and in such a manner. On the eleventh floor of the Cultural Education Center in Albany is a folio of colorful maps of the Hudson River, at a scale of 16 inches to a mile, covering the 16 or so miles from Troy to New Baltimore, with depths and sinuous shorelines. Also housed there is an eleven-map atlas of the Onondaga Salt Springs Reservation in central New York, soft or faded watercolors marking lots on the finely inked streets of Syracuse, Geddes, and nearby towns. In the Albany County Hall of Records a 23-foot map in eight parts traces the 154-mile route of the post road from Albany to New York City; taverns, breweries, milestones, tollgates, fish ponds, individual houses, and, naturally, post offices are noted in a clear hand. All these many maps are Randel's.

Randel's story has not been chronicled in detail because few of his letters or records have survived or surfaced. But there are some scattered archival materials. Taken together with documents from contemporaries and colleagues and historical scholarship, they shed light on Randel's life and work, on his marriages, his children, his religious feeling, his financial woes, his lawsuits, his anxiety about his reputation, his closeness with his family. His precision and exactitude won him a legacy but also heaps of trouble, and his long life was fraught with conflict and disappointment.

Randel was both emblematic of his time and a visionary well ahead of his time. He was of the Enlightenment, born into a culture and a period in which reason and measured action were prized and dominion over the natural world—through exploration, experiment, science, cartography, and infrastructure—was celebrated. His was the era of laying lines on the land—lines for communication, for transportation of people and goods; lines for establishing nationhood, statehood, and individual ownership. Those are the lines, the geometry, that define much of the American landscape today; they represent a particularly American relationship with land and identity. Randel also lived long enough to see the waning of some of the Enlightenment values he embraced and the ascension of the Romantic era. During his middle and old age, profound shifts in thinking about land, nature, and cities emerged; some of the most significant shifts took place in New York City and New York State, where Randel did much of his work.

Today, many ecologists and planners strive to lift the lines that people like Randel laid down. They do this to restore, to the fullest extent possible, the habitats of plants and animals—to put the natural curves back in canalized rivers, for instance, so the waters can flow as they once did. Planners and ecologists do this to ensure that ecosystems retain or regain some of their vibrancy, health, and well-being and that we retain and regain our connections to the natural world. They do this to try to ensure that our future is biologically diverse and sustainable, that cities become more livable. Many can do this only because Randel and surveyors like him kept meticulous, valuable records. Contemporary researchers rely on the data and accounts of surveyors to recreate past landscapes, and they rely on their maps. Randel's beautiful maps and his careful measurements readily interface with Geographical Information Systems of today, giving rise to creative and thought-provoking works.

Randel's measurements and maps have been central to two innovative projects. Rose-Redwood used Randel's data to explore whether Manhattan was flattened and to quantify the displacement the grid brought in its wake. His research into Randel's mentors and employers also deepened knowledge about the origin of the grid plan. Landscape ecologist Eric Sanderson of the Wildlife Conservation Society has used Randel's data to create a digital version and narrative of what the island might have looked like in 1609, when the first modern European visitors, Henry Hudson and the crew of the *Half Moon,* arrived. Sanderson's book *Mannahatta* and related online project, *Welikia*, have catalyzed new ways of seeing and thinking about how cities evolve and what they might become.

Randel was convinced he was ensuring a wonderful future for Americans and for the city of New York. By many measures, he did. There is continuity between the methods Randel used to lay down infrastructure in the nineteenth century and the emergence of the modern information infrastructure, the satellite system, which governs many aspects of modern life and of surveying. Nineteenth-century infrastructure has also become in some places its own antidote: old rail lines live as bike paths, trails through woods, or a garden wending within a cityscape. Canals live as meandering, recreational rivers. And even as the grid plan of 1811 softened the extremes of the island's topography, smoothing some undulations, it catalyzed a different kind of elevation: buildings grew tall; people who wanted to be on the vibrant island came to live densely. That density is leading New York City to grapple with its future in ways that resonate for people everywhere, for cities and urban areas now house more than half the world's population.

This story of the once and future city begins at a bolt on a rock in a park on the island.

II

In Which John Randel Jr. Affixes the City to the Island

John Randel Jr. encountered much vexation in the vicinity of 59th Street and the Fifth Avenue, as he called it. For about a decade, Randel knew 20 or so square miles of the island better than anyone alive. There were few rocks, fields, marshes, and thickets he did not repeatedly visit year after year, as he measured and remeasured, as he layered new data on top of old. His recorded visits to 59th Street and the Fifth Avenue, a somewhat swampy and distinctly rocky area, bespeak repeated frustration.

On one occasion Randel was irritated by his measures of block lengths. He was in the midst of placing iron bolts or official monuments—slabs of white marble, 3 feet high and 9 inches on each side—at more than a thousand future intersections, so avenues, streets, and rectangular blocks could be laid out correctly as development swept up the island. The north-south blocks were to be 260 feet long, but one of the markers he had in place at 59th Street was in the wrong spot. "This monument 59 I expect must be wrong though I cannot detect an error in the calculations. I have measured the line, which is level, twice and find it 260 feet and 65 hundredths instead of 260 feet and 80 hundredths. I therefore have altered the monument by moving it north 15 hundredths," he wrote in one of his field books.* Errors of such magnitude—the discrepancy was a full 1.8 inches in this case—tormented Randel.[1]

On another occasion near 59th Street and the Fifth Avenue, Randel accidently broke a screw on an instrument and had to travel to the city, 3 miles to the south, to have it repaired. He lost a day's work, which he could ill afford that season. It was the fall of 1812, and Randel's brother William, on whom he relied, had been ill much of the unusually cool summer, unable to survey, so fieldwork had slowed.

As winter arrived, Randel suffered further setbacks. "Rained all last night and continues raining . . . Sperry, Plamondon, Charles, Robert and Shannon attended to their instructions today though the storm was very severe. Mr. Wood and Clarkson refused and went to the tavern near Mr. Posts . . . and remained till night. This is the first time anyone has absolutely refused. The cause they give is the inclemency of the weather. The weather was severe, I admit. But while the others had to go and did go, and the monument they had to carry was necessary to be carried, it was not proper that they should be excepted while the others had to toil in the rain and the water," he wrote in his field book. The following day also promised rain, and Randel tried to prevent the two troublemakers from again absconding to a tavern. "Mr. Wood and Clarkson are to carry the monuments they neglected yesterday to their places. Charles is to go along . . . and see that time is not unnecessarily delayed and not to carry any himself unless one of them refuses . . . And I now promise Mr. Wood and Clarkson that the one refusing today or neglecting to carry as many as should be carried shall have $5 deducted from their wages and [be] discharged tonight."[2]

Life at home was no easier. Randel's rented house in the village of Harlem was not secure. He had corked the windows, purchased

---

* To make it easier to understand Randel's writing and mathematical notations, I have spelled out some abbreviations and added some punctuation.

a lock for the stable, and taken particular precautions to protect his horse by leaning a rail against a hole in the stable door and resting its base in a hole in the ground—using the principle behind the police locks that became common a century later in many Manhattan apartments. Even so, the horse got out. "Mr. Low told me he had my horse and would pound him for getting into his turnips unless I would pay him for the damage. I told him I had agreed to pay Mr. Port, my landlord, for all damage done in the lot and would pay him if he got Mr. Port's order for the money. He appears to show by his countenance that today was Parade Day. I do not think proper to indulge him, therefore refuse paying any damage unless he produced Port's order. He goes off Miffed." Escaped or loose domestic animals were often rounded up or boarded in the town pound; owners could claim them after paying the pound master.[3]

Taking care of pound business would not have been a priority for most New York City men on parade days, however, when many of them assembled at the southern end of the island to demonstrate military might with an assortment of weapons. In 1812, New York was particularly concerned about its defenses, because war had been declared against Great Britain in June. As the son of a Revolutionary War soldier, Randel knew the value of such military assemblies. But as an employer faced with an enormous task, he was not sympathetic to his men's absence. Military training days could have a country-fair quality, "a carnival atmosphere," explains Matthew Keagle, an expert on New York City's Seventh Regiment. Indeed, by the 1820s such parades were openly satirized in cartoons and articles. This parade day, three of Randel's men were absent so they could make a show of force and have fun, leaving him shorthanded in the field.

The next day, having reflected on the plight of the horse, Randel realized mischief was perhaps at play. "I direct McDonald to give the horse oats in case Low does not take the horse to pound tonight, and in the morning, in case the horse is taken, to request the Pound Master to give the horse 10 or 12 quarts of oats daily and take good care of him for me and he shall not come by it. Note: As it was improbable for the horse to get out of the stable alone, he must have been let out either by Port's men or Low, who had no business with my stable—who also must have taken down the partition fence between the yard and turnips that I had put up. They therefore are subject to pay damages, I think," he wrote. A battle of wills ensued, with Low and Port asking for damage and pound fees and Randel refusing. A little more than a week later, the horse was stolen: "The horse and chair harness is missing this morning. The gate and stable door that was closed by my brother last evening when we returned from town is open early this morning (6 o'clock)." Randel sent one man east and north to check Hell Gate, Harlem, and Kingsbridge, another to Brooklyn, and his brother William south and west to the Hoboken and Paulus Hook ferries to look for the horse. That night, the horse was found in a Mr. Peter's shed.[4]

Mishap and misadventure dogged many nineteenth-century American surveyors. Crews were unruly and in flux. The weather could impede and endanger a survey team. Instruments were scarce and imperfect. Injury and illness, even death, were common. Surveyors encroached on Native American territory and were sometimes under threat of attack. The landscapes they measured and assessed were often remote wilderness, alive with top predators, far from supplies and comforts. As one nineteenth-century surveyor put it, "When the heavens open and the rains descend, the settler enters his log-cabin, unpretending as it is, and he is secure from the beating storm. Under similar circumstances, the surveyor has no alternative but to lay his troubled head on a soft lightwood knot, his body on the wet ground, and let it rain, thanking his stars it isn't a hail-storm."[5]

Randel faced almost all these challenges. His team was always reconfiguring—some men were fired for drunkenness, gambling (in Mr. Low's gambling house, which Randel ultimately reported to the police), or laziness. Fits of rage and anger were not uncommon. "I James Wilson acknowledge that I have wrongfully and without cause attempted to injure the character of John Randel Jr. because I was dissatisfied with his discharging me notwithstanding he paid me one dollar more than I agreed to work for him for. I now in the presence of his men and himself beg the pardon of said John Randel Jun for said ill treatment." Wilson's confession and apology appear in one of Randel's field books. But such testimonials are rare—many men did not redeem themselves. Randel was repeatedly robbed. Hard winters and the rain and the wind frequently delayed him. He suffered a litany of illnesses and injuries. His instruments broke or were no match for the island's topography, and so Randel ultimately devised his own.[6]

Randel encountered wilderness when he worked upstate, but not in Manhattan. Although the island was far from fully settled, it was no wild land. Europeans had colonized it two hundred years before, and Native Americans had occupied it for more than 11,000 years. Paths such as the Eastern Post Road, Bloomingdale Road, and Kingsbridge Road laced the island from end to end, often following the routes of earlier footpaths, and country estates and farms were plentiful. The terrain was rough and vegetation thick in many places, but wolves no longer roamed the island, and cougars and black bears were well gone. During their occupation between 1776 and 1783, British troops had razed the ancient forests of the northern part of Manhattan for fuel and fortifications; only saplings would have regrown by Randel's time. And Randel was not under attack by Native Americans; the Lenape had in large part left the island by the early nineteenth century. Rather, Randel's attackers were farmers and property owners, and sometimes their dogs.

Perhaps the greatest challenge Randel faced—the thing that cost him time, money, and frequently his health—was contending with his own desire to get everything perfect, to get every measurement, every marker, every map exactly as it should be. Randel did not merely survey Manhattan as any land surveyor of that time would have. He brought remarkable sophistication to his work, sophistication unusual for someone working on a survey that no government—city, state, or national—had deemed scientifically or strategically significant. Working alone, following his heart and mind, Randel came to practice what is called geodetic surveying. It put him on a par with men now famous for such work, such as Charles Mason and Jeremiah Dixon, Swiss mathematician and surveyor Ferdinand Hassler, and Scottish military surveyor William Roy.

It took Randel twelve years to complete his work on various aspects of the Manhattan grid. When he started, he was an unknown twenty-year-old whose biggest jobs had been drafting maps based on field reports, surveying turnpikes, and measuring property lots in Albany and central New York State—hardly remarkable work for a young surveyor of that era. By the time he was finished with the grid, he was recognized nationally for his talents and was able to make the transition from surveyor to the more prestigious emerging profession of civil engineer. He had commissioned a famous painter to make portraits of himself and his family. He had a reputation. He had arrived.

Randel had not been the city's first choice for the job of surveying the island. In 1806 the Common Council—the city's governing body, which included the mayor as well as elected aldermen from each of the city's then nine districts or wards—had offered such a job to Ferdinand Hassler, who had recently emigrated from Switzerland and was living in Philadelphia. European men of science were often perceived, often correctly, as more technically expert and experienced than their American counterparts. (Although at times they were also overlooked for just that reason. As a New York canal commissioner wrote, "I would recommend employing Americans solely, and avoiding foreigners . . . the truth is the laying out a path for a canal neither requires conjurers nor wizzards;—practical nature is every thing that is necessary.") Fortunately for Randel, Hassler was not hired. The explanation for the subsequent appointment of a young unknown over many much more experienced men, and particularly many more New York City–savvy surveyors, lay 150 miles to the north, in Albany, with a powerful man named Simeon DeWitt.[7]

To the Place of Beginning

Randel was born in Albany in 1787 to a large, close family that helped him in many ways with their talents and skills and that he in turn loved and supported, often to his financial detriment. His father, who lived from 1755 to 1823, was a brass founder and jeweler whose own father had come to America from Randeltown in Antrim County, Ireland. His mother, whose first name is alternatively spelled Catherine, Katurah, or Keturah in church and genealogical records, and who lived from 1761 to 1836, was born in New Jersey to Jesse and Rebecca Fairchild. She was part of an extensive family of New Jersey first settlers, a land-rich family central to her son John's life. During the Revolutionary War, Randel Sr. served as a private in New York's Second Regiment under Abraham Riker, Philip Van Cortlandt, Nicholas Fish, Peter Schuyler, John L. Hardenbergh, and Levi DeWitt, officers whose connections seem to have benefited the Randels. Fish became an alderman for New York City and oversaw Randel Jr.'s Manhattan work; and Levi DeWitt, who later stayed with Randel Jr. in New York City, linked the Randels to his influential Albany family. Of Randel Sr.'s day-to-day service, only one detail is known: He served on the second night watch for the Powder House on April 28, 1775. The watchword was "Pitt" and the evening uneventful, although the week had been anything but. A group of New Yorkers, "the Committee of Sixty," had called for a stronger revolutionary government, one with the power to direct the war effort. And the city had received news about the battles of Lexington and Concord, the first of the war; panic and revolutionary insurrection had ensued, with New Yorkers raiding the city arsenal.[8]

Sometime after the fighting reached New York, Randel Sr. was captured by British forces and sent to Halifax, Nova Scotia, where he was imprisoned. Conditions in the jails and prison ships were brutal. An estimated 18,000 prisoners died of starvation or the diseases endemic to filthy water and inhumanely crowded quarters awash in excrement and vermin. Despite the hardships, Randel Sr. was keenly observant of his surroundings, later describing in detail the harbor and interesting vessels to his son John.

Randel Sr. survived Halifax and was discharged in 1779 as a casualty. It is unclear whether he and Catherine Fairchild met before the war or after he was freed, but they married a year after he left the regiment and settled in Albany, first staying in the boarding house of Paul Hogstrasser, perhaps while Randel Sr. established his business. A direct descendant has Catherine's wedding ring—a thin gold band set with a small ruby, decorated with an intaglio of swirls and engraved on the inside "J. R. to C. F. 1780," likely made by Randel Sr. himself. Catherine bore many children. One record says eleven, but clear documentation exists only for nine: in order of birth, Hannah, Rebecca, Daniel, Abraham, John, Jesse, William Stevens, Sally Eliza, and Jane DeWitt. Two others, Catherine and Charlotte, are mentioned as sisters in Randel's field books. A sister Mary appears in other records. One brother, Samuel Dodd, lived only six weeks; infant mortality ranged between 10 and 30 percent and could be higher during the frequent outbreaks of measles, yellow fever, smallpox, and cholera. The few surviving Randel family letters capture the siblings' fondness for one another and their parents, their humor, their diligence and their love of music, their anxiety about stretches when they have not had news of each other.

ALBANY WAS BOOMING at the turn of the nineteenth century. The Randels were among the 9,000 or so people who settled there between 1790 and 1800, and like many of the Scotch-Irish immigrants, the Randels belonged to the First Presbyterian Church, a congregation vividly described in David G. Hackett's *The Rude Hand of Innovation*. The Presbyterian church was the choice of newcomers unwelcome in the well-established, wealthy Dutch church—although even for newcomers admission was notoriously difficult, because the ministers relentlessly scrutinized every aspect of applicants' and members' lives. Immigrants more readily assured a place for themselves in the community if they

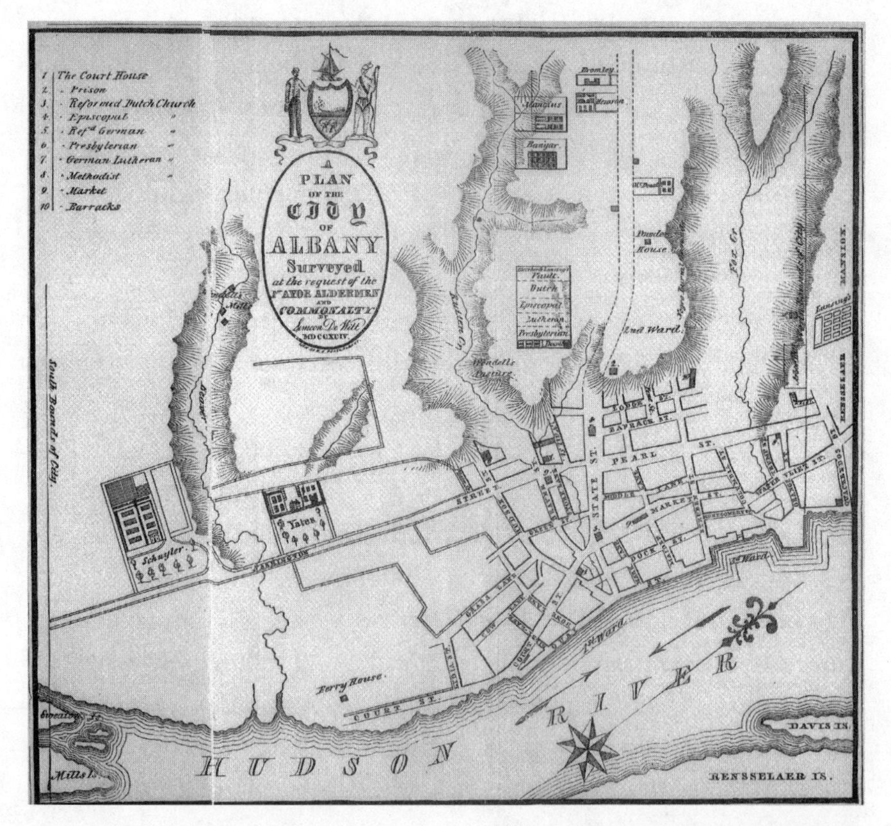

1794 map of Albany by Simeon DeWitt. Courtesy of the Albany County Hall of Records Archival Collection.

were able to join the church, and Hackett attributes this success to the rigor of the Presbyterian ministers: "The extreme discipline needed to manage one's economic affairs in order to survive as a stranger in a hostile town paralleled the rigid moral discipline demanded for one's spiritual salvation."[9]

The church molded Randel Jr. His experiences in the Albany Presbyterian congregation shaped his approach to business, to his employees, family, acquaintances, rivals, reputation, and to hard drink. Respectful references to church and sermons can be found throughout his field books. Even in a remote, newly established town in central New York, in the midst of a surveying trip and severe snowstorm, Randel stopped to assist in the dedication of a First Presbyterian Church. He often noted which minister he heard on which Sunday, which passages they read; one of his field books included a prayer. A young man working for Randel explained (or complained) to his mother that he was unable to send many letters due to "Mr. Randel being conscientiously scrupulous respecting our breaking the Sabbath by writing."[10]

Randel in later life spent enormous energy and money defending his reputation, and it is easy to understand why reputation so obsessed him. When he was a boy and a young man, the elders of the First Presbyterian Church visited families once a month, or more frequently, exhorting them to behave and to work hard in their public and private lives. The elders would then meet and discuss each family. Twice a year, the worthy would be invited to take communion. The uninvited became suspect to their peers, their reputations questioned; they entered what Hackett describes as "a purgatory of public opinion." Sometimes the elders appointed a committee to bring the fallen "to repentance." Sometimes they chronicled the sins of an individual or a family before the entire congregation. Whether the Randels were ever shamed in this way is not known. But their culture was one of constant surveillance and harsh judgment. Everyone knew everyone else's moral business.[11]

While extolling hard work and honesty and exacting good behavior (and donations and pew fees) from their flock, the church elders also promoted education. When Randel was thirteen, the community's already strong emphasis on schooling became focused more specifically on science, with the arrival of minister Eliphalet Nott. "His ambition was to make men *wiser* and *better*, rather than to promote the sectarian interests and speculative tenets of the church . . . All I mean to say is, that Doctor Nott was by far the most eloquent and effective preacher of the period to which I refer," wrote a contemporary. Making men wiser and better meant embracing Enlightenment insights from natural science and philosophy, as Hackett writes: "Rather than turn away from these new ideas, which to some threatened to lead Christians away from the Bible's truth, Nott held that the new knowledge was not opposed to the revelation but deepened one's knowledge of it." Nott's sermons—and Randel surely heard many of them during his early teens, and recorded his attendance at several during his twenties—included references to Isaac Newton's discoveries, to John Locke's philosophy, to the vital role of science, and to the need for education in science. Randel followed this teaching, embracing science and new technologies all his life.[12]

Randel attended a primary school of fifty to eighty pupils run by Mr. Ermis (or Eumis), who taught him the first principles of mathematics. Randel excelled in math—one acquaintance described him as an "eminent mathematician"—and Ermis was influential for him. "I have always esteemed him a mathematician of the first class . . . he has a particular faculty in teaching children," Randel wrote in an 1820 letter of introduction for Ermis. "I am personally very much attached to him." If Randel went to college, it was not Harvard, Brown, Yale, Union College, Columbia (then Kings), Princeton (then the College of New Jersey), Rutgers (then Queens), or the United States Military Academy at West Point (established in 1802). He was too old for the Albany Academy, which was created in 1813. However his education was procured, it was excellent. His penmanship was stunning, his writing beautiful and descriptive. He knew physics and mechanics in addition to math and was versed in astronomy. In one of his field books he cited William Enfield's *The History of Philosophy*. In another he drew a map of the world from Samuel Vince's *A Complete System of Astronomy*, which he attempted to teach his younger brother William. Randel set himself challenges as well. James E. Morrison, a scholar of star maps called astrolabes that were used for timetelling and navigation, has examined some of Randel's calculations and concluded that he was trying to teach himself to use an astrolabe. "His use of spherical trigonometry is dead on," Morrison says, but Randel gave up on those calculations, "possibly because he realized he was doing it wrong." That failed attempt aside, Morrison says, "Randel was clearly very bright, well educated and meticulous." He may even have achieved some notoriety for his meticulous memory, which served him well when doing math in the field. Randel seems to have been invited to give two lectures on mnemonics in New York City in 1815. Based on descriptions from his youth that he provided when he was in his seventies, his memory did appear to be what would come to be called photographic.[13]

As an older man, Randel stayed on top of developments by reading and by keeping up his membership in professional and scientific organizations. At various times he belonged to the Franklin Institute in Philadelphia, the American Institute of the City of New York, and the American Association for the Advancement of Science, where lectures by members and visitors disseminated discoveries—and for which Randel sometimes undertook experiments and wrote reports. In his forties he owned a copy of *The Elements of Mechanics* by James Renwick, a professor at Columbia College; his copy now belongs to a gentleman in Maryland. Although only that one book has come to light, Randel likely maintained an extensive library. He was known to the poet and traveling bookseller Nathan Lanesford Foster. In 1840 Foster tried to sell a book to Randel in Philadelphia at his brother-in-law's house, but "his wife would not take it," Foster wrote ruefully. "Went without dinner." Foster then sought to deliver the book to Randel in Wilmington.[14]

Reading, setting himself assignments and problems, and attending public talks may have been the main means through which Randel was educated as a youth as well. Albany was alive with educational discourse when he was growing up, and some of his learning must have come from exposure to lectures and discussion there. The Albany Society for the Promotion of Useful Arts, founded in 1791 as the Society for the Promotion of Agriculture, Arts, and Manufactures, and to which Randel belonged, held and published talks on such topics as unifying the system of weights and measures, improving carriage springs, observing eclipses, and deterring wolf predation (by perfuming the sheep with offensive scents, an approach adopted by some western ranchers more than a century and a half later).

The Albany Mechanics Society, which Randel Sr. helped found, provided instruction as well. Craftsmen and tradesmen formed the Mechanics Society in 1793 to aid families of members who became sick or injured and to provide schools for their children as well as for indigent children. In 1811 the society established a bank for its members, which Randel used for financial transactions. The interests of the Mechanics Society reached beyond establishing financial security and a supportive community. Many members viewed innovation, specifically mechanical power and devices, as protection against life lived as "a savage in the wilds of nature." They urged self-improvement and technical knowledge, which was the focus of the education they supported. As Hackett writes, "The biographies of Albany craftsmen suggested they were reared in this ethos. Unanimously described as ambitious, most of the town's successful craftsmen left their homes as young men, frequently chose an occupation different from that

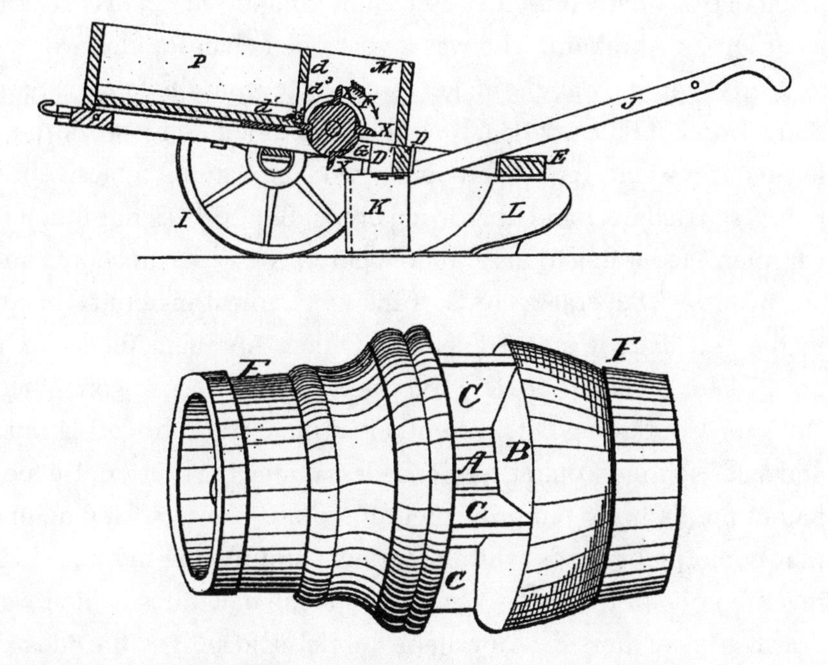

Two inventions by Abraham Randel: "Improvement in Potato-Planters" (top) and "Hub" (bottom). United States Patent and Trademark Office.

of their fathers, and often bristled under the temporal and paternal restraints of an apprenticeship." Randel doesn't seem to have bridled under the tutelage of older men he respected. Neither his father, who aided him professionally, nor his mentor appear to have elicited rebelliousness or chafing against authority. Randel did outspokenly resist and rebuke older men whom he did not respect or who he felt did not (or could not) discern truths made clear (to him) by math, science, and pragmatic thinking.[15]

Randel Sr.'s sons did not follow him into brasswork or jewelry making, but some did follow him into craftsmanship. William, who frequently assisted his older brother with surveying, made cabinets and furniture: ". . . a large assortment of MAHOGANY AND CHERRY FURNITURE, consisting of elegant Side-Boards, Secretaries, and Book Cases," read one of his advertisements. Daniel was listed in the Albany directory as a cabinetmaker; Jesse was listed as a carpenter. Abraham, who was a year older than Randel, worked as a surveyor and cabinetmaker; an invoice from the famous Sans Souci Hotel in Ballston Spa, New York, documented his work there for about a week, repairing popular wooden gaming tables. Abraham also tried his hand at growing crops. But he "was not much of a farmer," according to his granddaughter, "he was a mechanic and an inventor." Several records of his mechanical ingenuity reside in the U.S. Patent and Trademark Office. Abraham filed patent no. 374 for a "Mode of Constructing Carriage and Wagon Wheel Hubs for Containing Oil," essentially a means of lubricating hubs and axles so they would experience less damaging friction. He filed patent no. 5446 for "Improvement in Potato-Planters," a machine that could plant crops while dropping "stimulating manure." He invented one of the first successful mowing machines, which was drawn by two horses. "Abraham Randel said he got the idea of the machine from watching his wife using a pair of scissors," his granddaughter recounted. The machine was later manufactured, and Abraham received $600 (about $15,000 today) for his design. Patent experts B. Zorina Kahn and Kenneth L. Sokoloff note that inventiveness facilitated survival in the early American republic. People like the Randels had to rely on themselves to solve technical challenges, and most early patents originated with regular people, Kahn and Sokoloff find, not with "an elite with rare technical expertise or extensive financial resources." Inventiveness was also considered, or has come to be considered, a particularly Yankee trait. The Randel men certainly had it.[16]

The Randel women may well have had it too. They had lively characters and intelligence; their few surviving letters are suffused with curiosity, insight, and warmth. Randel's mother, Catherine, maintained a dynamic correspondence with her younger relatives. A grandson wrote, "Dear Grandma, I would like to see you very much," and went on to describe his mother's "13 different kinds of geraniums, and a rose bushes that have got roses on, one down in the kitchen and the other up in the study." Catherine desired particulars about all her children's activities. "Mother says she wishes you would write by Mr. L and tell us all about your crops," wrote her daughter and namesake, Catherine, to her brother Abraham. "If your corn grows, if your peas, Beans, cucumbers, pumpkins, squashes, melons, wheat, rye, Barley, Oats, hops and every thing Else you have in the ground grows or is likely to grow."[17]

This Catherine, Randel's sister, had a terrific sense of humor. Admonishing her brother and sister-in-law Rebecca for failing to correspond more often, she wrote, "We have not heard from you since sister H[annah] was here, except once brother john mentioned in one of his letters receiving one from Brother and I feel as if you and Brother A might make out to write a letter between you each writing two lines a day, in the course of a week you would have almost a page and a half wich would make a tolerable letter, and we would think much of it."[18]

In the same letter Catherine hungered for news, a common theme, separated as the family was for long stretches of time and often by many days' journey. She wanted to know everything: how people were traveling and for how long and when and how long they would stay in various places and "all the little occurences of the day." She described her days in Bloomfield, New Jersey, with wit and perspicacity. To resolve a family debate about whether Cousin Jeptha had recently lost his color, she watched Jeptha play the dulcimer and flageolet one afternoon: "I stood almost behind him, not to confuse the gentlemen, and had an opertunity of noticeing if he had lost his collour, and from the observations I made while standing, I thought the rose and lily were equally blended in his cheeks."

Catherine wrote of the night as well: "For three nights past, rather north of west we have seen a comet with a tail two yards long, what that prognosticates I shall leave you to determine, some say it is a sign of war." Across the Atlantic, astronomer William Herschel wrote to his sister and fellow astronomer Caroline of that same comet: "Lina:—There is a great comet. I want you to assist me. Come to dine and spend the day here. If you can come soon after one o'clock, we shall have time to prepare maps and telescopes. I saw its situation last night. It has a long tail." The great comet that entranced both the unknown Catherine Randel and the famous Herschels was ultimately named for Johann Georg Tralles, a German surveyor, physicist, and mathematician who had instructed and worked with Ferdinand Hassler. The two collaborated on a survey of Switzerland shortly before Hassler left for America with the intention of becoming a farmer, shortly before New York City officials sought to hire him to survey Manhattan.[19]

An Art That All Mankind Know They Cannot Live Peaceably Without

Perhaps because of his obvious facility with math, Randel found his way into surveying. Such jobs were plentiful. Randel grew up during America's surveying boom, a time in which distinctly American approaches to the craft emerged in response to the country's political and social needs and in response to its terrain. In Randel's time, many young men went into surveying—temporarily, at least. "I have a theory that in the nineteenth century, every middle-class white man in America surveyed at one time in his life," says Deborah J. Warner, a curator at the National Museum of American History and an expert on surveying history. Many now famous men served stints as surveyors, from early presidents—George Washington, Thomas Jefferson, John Adams, Abraham Lincoln—to literary figures such as Henry David Thoreau. The American boom resembled England's during the mid-sixteenth century, when the crown seized Catholic Church holdings: the church lands needed to be circumscribed and described before they could be sold.[20]

As Andro Linklater notes in *Measuring America*, and as other historians of surveying document, the transfer of property from church to crown to landowner reconfigured the British approach to land and to surveying. Precise measurement became crucial to establishing values for rents or for crop yields, and surveying flourished as a profession. Mathematicians and surveyors devised techniques and equipment, including the ingenious Gunter's chain. The chain, as Linklater describes so well, was remarkably versatile and clever. It was 66 feet long—equivalent to 22 yards or 4 rods, which were also sometimes called perches or poles—and divided into 100 links. Edmund Gunter's 1620s invention enabled surveyors to convert easily from rods or perches into feet or yards. Yards in turn could easily become acres: one acre is 4,840 square yards, the same as 10 square chains; 640 acres make one square mile. Many how-to texts describing the use of Gunter's chain and other instruments and surveying strategies were published in the seventeenth and eighteenth centuries.[21]

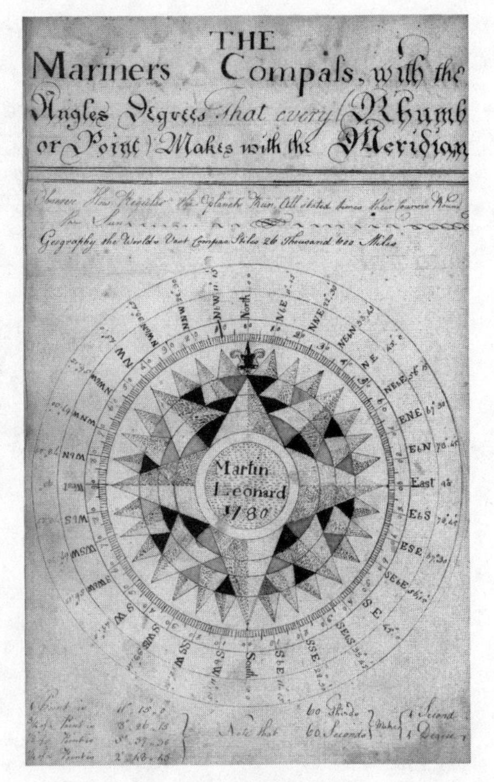

Student's surveying workbook from 1780. Courtesy of Jeffrey Lock, Colonial Instruments.

Those texts were plentiful in Colonial and Revolutionary America, where the parceling and sale of Native American land was a principal national activity. John Love's *Geodaesia: or, the Art of Surveying and Measuring Land Made Easy*, first printed

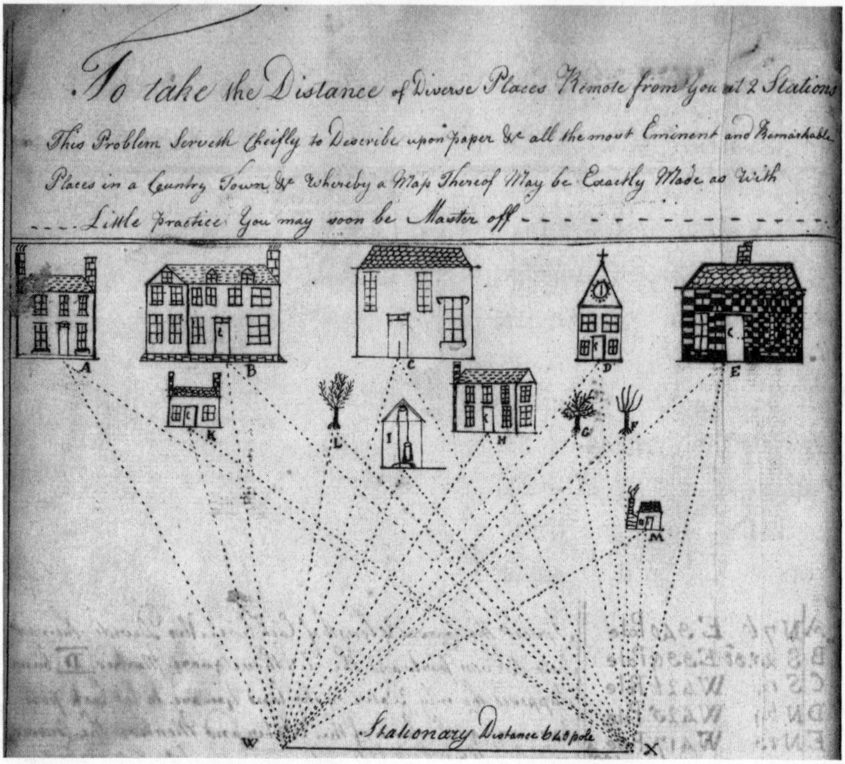

Student's surveying workbook from 1830. Courtesy of Jeffrey Lock, Colonial Instruments.

in London in 1688, was one of the most popular such texts during Randel's childhood, when it entered its thirteenth edition; George Washington apparently relied on it. Love celebrated surveying: "It would be ridiculous, to go about to praise an art that all mankind know they cannot live peaceably without, and is near hand as ancient (no doubt on it) as the world: for how could men set down to plant, without knowing some distinction and bounds of their land?" And Love made clear that part of his mission in laying out the "wholesome" principles of geometry, or geodesia, was to help surveyors on the other side of the Atlantic: "I have seen young men in America so often at a loss, that their books would not help them forward; (particularly in Carolina,) about laying out lands." Love's book set out step-by-step procedures and was filled with examples and problems. John Gummere's *A Treatise on Surveying*, a popular American book, did the same. Workbooks surviving from Randel's day show many students soldiering through Gummere's problem sets, some with wonderful care and flair (adding idiosyncratic details such as drawings of fish and men in top hats), some clearly caring less.[22]

Men wanting to learn surveying could study texts like Gummere's or Love's or could learn the essential math in elementary school, in college, or during popular short courses taught on evenings and weekends. Many learned on the job as apprentices, which was how Randel learned. He had the good fortune to study with Simeon DeWitt, who was for fifty years the surveyor general of New York State. For several decades on both sides of the turn of the century, DeWitt was among the most influential men in Albany and in New York.

I Have Wore a Piece off My Toe in Walking Too Much

DeWitt was born in Wawarsing, New York, in 1756 to a wealthy, well-connected family. He learned surveying from his uncle, James Clinton, the father of DeWitt Clinton, who was a U.S. senator, mayor of New York City, governor of New York State, and one of the main forces behind the Erie Canal. James's brother George was a New York State governor and U.S. vice president. Both uncles delighted in DeWitt and James recommended him to Robert Erskine, the Scottish engineer whom George Washington appointed in 1777 as surveyor general of the Continental Army. Erskine was a polymath. During the war he not only produced strategically significant maps but developed a plan for an iron fence, the Tetrahedron, that could run undetected underwater across a river to rupture enemy vessels. Erskine's idea was in the air: chains were stretched across the Hudson River during the Revolutionary War, and Erskine, who managed an iron foundry, forged some links for those chains. (The idea didn't lack creativity, only efficacy.) Erskine wanted to work with likeminded men, with "Young gentlemen of Mathematical genius, who are acquainted with the principles of Geometry, and who have a taste for drawing." He was offended when recruits did not meet his standards. "This is surveyed by Mr. Lodge . . . A Most abominably Lazy Slovenly Performance not to Survey such a small piece over again or lay it down properly," he commented in the corner of a New Jersey map.[23]

In DeWitt, Erskine found a true collaborator. DeWitt was well schooled (the only member of the Queens College—now Rutgers University—graduating class of 1776), deeply interested in math and astronomy, good at mapmaking, and inventive. He devised a rain gauge and a protractor to help surveyors do mapping in the field, and he designed one of the most detailed and delicate astrolabes of his time. (That astrolabe, housed in the Smithsonian Institution, is drawn and painted on paper and is quite small, only 5 inches on each side. It portrays the night sky at 41° N, the latitude of a New Jersey estate where DeWitt stayed in 1780.) During the war DeWitt worked on surveys of roads and terrain. "We are now in Quarters at New Windsor & live very well, but I Would Prefer the North Branch nevertheless because there are a Parcel of Clever fellows there besides we are here in A Wild woody rocky hilly country," he wrote to his college friend John Bogart in 1779. "I have been cursing the mountains ever since we came to this place because they tire me so much in travelling over them We have Survey'd nothing but the Paths and Passes on them—I have wore a piece off my toe in Walking too much it smarted yesterday very much but to day he Seems well enough." DeWitt's letters to Bogart are lively with humor and thoughts about pretty girls. Bogart, then living in North Branch, was decidedly among DeWitt's "Parcel of Clever fellows."[24]

Simeon DeWitt, c. 1804, by Ezra Ames (American, 1768–1836). Oil on canvas 153.7 x 123.2 cm (60½ x 48½ in.) Collection Zimmerli Art Museum at Rutgers University. Gift of the grandchildren of Simeon DeWitt. Photograph by Jack Abraham.

After Erskine's death in 1780, Washington appointed DeWitt surveyor general for the army, a position he held until 1784. "I can assure you, he is extremely modest, sensible, sober, discreet, and deserving of favors," Washington wrote to Thomas Jefferson. More than a decade later, President Washington asked DeWitt to become surveyor general for the country, but DeWitt was reluctant. "A man of profound knowledge in mathematics and sufficiently versed in astronomy, was nominated to that office, and has declined the acceptance of it," Washington wrote with seeming regret. At the time of Washington's offer, DeWitt had been surveyor general of New York State for a dozen years. He had become active in Albany's scientific and educational projects and was not eager to leave. He remained surveyor general until his death in 1834.[25]

DeWitt was responsible for surveying, mapping, and selling state lands and for giving Revolutionary War veterans acreage they had been promised in reward for military service. In order to sell land, the state government needed to own it. New York State's appropriations of Native American land were not as consistently violent as were such appropriations in the West. But they were unrelenting, and in the end just as ruthlessly effective. Through a combination of legislation, paltry purchases, treatises, and leases, New York appropriated almost all the Iroquois or Six

Nations land throughout the state (the land of the Mohawk, Seneca, Cayuga, Onondaga, Oneida, and Tuscarora). For instance, the New Military Tract—twenty-eight townships to be settled or sold by veterans—encompassed more than 1.5-million acres of Onondaga and Cayuga territory. DeWitt and many of his friends and associates, including Randel, did well by their insider knowledge of the land and its early value. Ithaca was DeWitt's particular project. Oneida County was Randel's.[26]

DeWitt's office surveyed, distributed, and sold thousands of acres in the New Military Tract and elsewhere to thousands of land-hungry, eager settlers. Settlement proceeded with such rapidity and fervor that one wave of it was called Genesee Fever, for the lands stretching between the Genesee River, which runs roughly north-south between Lake Ontario and Ulysses Township in Pennsylvania, and Lake Erie, the western boundary of New York State. (The other Genesee Fever frequently referred to in nineteenth-century accounts was probably malaria, typhus, or typhoid—perhaps all three. The Genesee region, at least in the summer, was moist and fertile, its swamps fetid, mosquito-filled.) An account written in 1795 described a single winter day in which 500 sleds traveled through Albany heading west; within three days, 1,200 sleds had gone through town, moving settlers into the middle of the state, toward the promise of land. Randel would have been seven that winter, and must have seen oxen straining to pull household-laden sleds across the frozen Hudson River and through the snow.[27]

Surveyors were the instruments of New York's transformation, as they were in every state. DeWitt brought rigor to New York State's approach to surveying and parceling land. Under the traditional method of surveying, called metes and bounds, surveyors would essentially follow natural boundaries (bounds) such as streams or ridges and would measure (mete) distances as best they could—plots could look like puzzle pieces. Parcels often followed the contours of the land and its natural features. DeWitt was one of the few surveyors in the original states to distribute rectangular and square parcels and to require his surveyors to use cardinal points. New York State's approach reflected the country's new land division system, established by Congress in 1785. The "Ordinance for Ascertaining the Mode of Disposing of Lands in the Western Territory" imposed a uniform north-south, east-west grid—a grid unconcerned with natural boundaries and topography, not because those drafting it were necessarily anti-topography but because the grid was, on paper, practical.[28]

The grids dividing the land west of the original thirteen states into 6-by-6-mile townships, and those in turn into 36 one-mile-square subsections, remain obvious from the air and from the ground. Visitors driving to glimpse sandhill cranes along the meandering, curving bed of the Platte River in spring must swing through endless 90-degree left- or right-hand turns. Nebraska farmland is packaged in squares and rectangles; the roads run straight to dead-end in perfect T's or to pass through perfect crosses, every angle right. The same grid undergirds the farmland of Wisconsin, Iowa, and Minnesota. Contiguous rectangular grids cover 69 percent of the land in forty-eight states, scholar Hildegard Binder Johnson notes in *Order Upon the Land*. The grids were often imperfect, because the north-south lines of longitude, or meridians, as they are called, converged as surveyors traveled north on the sphere of the earth. A northern boundary could be 30 to 40 feet shorter than the southern line. ("Shrewd speculators learned to avoid land in the northwest corner of a township, because any errors in measurement showed up there, and that was where most arguments over boundaries occurred," observes Linklater.)[29]

Although imperfect in practice, the clean, clear lines gave rise to descriptions easy to understand financially, legally, politically. The lines mapped beautifully from afar, providing pioneers with clear direction. "The less men knew about the land, the straighter were the lines they drew—right across rivers and mountains," Johnson writes. The grid also embodied the spirit of the age. "The philosophy of the seventeenth century glorified *l'esprit géométrique* of the Cartesians. Thomas Jefferson and other eighteenth-century Americans were dedicated to a rational approach to the problems of their day," she notes.[30]

DeWitt embraced *l'esprit géométrique*. His surveyors were trained to lay lines at right angles, to be precise and exacting. DeWitt was not casual about surveying, as some surveyors were—even some directors of the federal land ordinance office. He issued obsessive-sounding directives. Excerpts from an undated letter DeWitt sent to his deputy surveyors give a sense of that instruction and the care DeWitt took to prevent some standard pitfalls:

> The first thing to be done by surveyors is to provide themselves with the proper Instruments, to wit a compass, Chain, paper, field books pens, Inkstand, ink phial, protracting Instruments and a marking Iron. No compass should ever be employed that has not been proved to be good. The Chain should be accurately measured and adjusted not only before using it but at least once a week while it is in use, and the surveyor ought always to be provided with an accurate measure for the purpose. [Temperature expanded or contracted metal chains.] The Surveyor should always have a Flag staff to set for his object and not trust to trees or other objects unless they are so remarkable that they cannot possibly be mistaken for others. Backsights should be taken as often as possible to ascertain whether the needle is not affected by mineral attraction or any other cause. [Backsighting, a way of checking one's work, is explained below.] No muskets, axes or any other thing that will attract the needle must be proffered near the Compass while a course is taken . . . Every line must be marked so as that it may be easily followed through the woods, for which purpose a sufficient number of trees along each side of the line must be marked with a blaze facing the line and one on each side in the direction of the line and all trees standing in the line must be marked with

Detail of the 1804 map of New York State by Simeon DeWitt. Courtesy of the David Rumsey Map Collection, www.davidrumsey.com.

> three notches and a blaze on two sides in the direction of the line. The beginning of every lot must be marked in a particular manner and a post or other measurement fixed at every corner and the nearest found tree to it marked and its bearing and distance noted in the description.[31]

DeWitt's implementation of the grid made New York State property speculation straightforward, just as in the West. The settlers and the grid they populated also permanently disrupted what remained of Native American ways of life, although any boundary lines, orthogonal or not, would have achieved the same effect. Land was generally not owned by individuals, according to Native American views; it was shared. The late eighteenth-century land division laws "required that reservation land be subdivided, with individual parcels allotted for each family. Plotting of individual lots for families undermined the traditional concept of land held in common and the solidarity of tribal control," writes Jo Margaret Mano, a geographer at the State University of New York at New Paltz and an expert on DeWitt. She describes the lines as "a net of settlement"; boundaries and fences restricted movement, and with less land, tribes were less able to sustain traditional hunting and harvesting.[32]

Just as DeWitt's administration of the grid supported the state's political and economic ends, so too did his cartographic style. The American maps promoting development were similar to European maps of the New World in that they depicted empty spaces ripe for colonizing, featureless landscapes easy to settle. In DeWitt's 1802 New York State map, such a story could be read. The map showed boundaries, some gridded and subdivided regions, some landowners' names, river routes and roads, some unimposing topography in the southern part of the state (but no disruptive Adirondack Mountains in the north), some water depths, some swamps. No icons depicted Indian lands, as some earlier maps had. And a great deal of land appeared empty. Those blank spaces, Mano argues, suggested land yearning to become productive. By the time the next comprehensive New York State map appeared, David H. Burr and DeWitt's atlas of 1830, the state was a study in grid. Tent icons appeared, but only to the far north: "The map gives the message that all the Indians had left for Canada," Mano writes.[33]

Blankness in American maps piqued Henry David Thoreau, who remarked on the distance between the land as it was and the land as portrayed. "How little there is on an ordinary map!" he wrote in his diary. "How little, I mean, that concerns the walker and the lover of nature. Between those lines indicating roads is a plain blank space in the form of a square or triangle or polygon or segment of a circle, and there is naught to distinguish this from another area of similar size and form. Yet the one may be covered, in fact, with a primitive oak wood, like that of Boxboro, waving and creaking in the wind, such as may make the reputation of a county, while the other is a stretching plain with scarcely a tree on it. The wauling woods, the dells and glades and green banks and smiling fields, the huge boulders, etc., etc., are not on the map, nor to be inferred from the map."[34]

It is, of course, impossible to know what most people imagined when they looked at one of DeWitt's maps, with its patches of featureless background. Some of the most evocative maps are the most bare; the viewer projects on the map what she or he imagines or needs to imagine. Polynesians using maps of the Marshall Islands in the South Pacific anticipated the tug of currents, the swell and pitch of waves, as they looked at sticks, shells, and fibers bound together in patterns. Maps of imaginary realms, such as C. S. Lewis's Narnia or J. R. R. Tolkien's Middle-earth, have few details, yet they conjure textures, feelings, distances, and landscapes that can seem more real than the landscape one stands in. DeWitt's maps accompanied rumors of lush lands in western New York, and perhaps many people's minds filled in thick forests soon to be fields.[35]

More and more Americans came to look at, own, and bring their imaginations to bear on maps during the nineteenth century, as a robust market in maps and geographic texts emerged. The dissemination of maps formalized ways of seeing land. Maps depicted in tangible fashion ownership, progress, and the impression of unity—essential ingredients for national identity. That individuals or households might own a map, once the rare and sole province of merchants, explorers, and royalty, reflected the new nation's principles and fostered pioneering fervor. One of the men instrumental to publishing and disseminating these geographies and maps was Mathew Carey, another powerful figure who was to play an important role in Randel's life. But that was to come later, well after Randel met DeWitt.

Bearings

DeWitt liked to hire young men whose families he knew. He may have come to know Randel's father and family through various Albany institutions, and because DeWitt's brother, Levi, had served with Randel Sr. during the Revolutionary War. The Randels valued their association with the DeWitt family, honoring them by naming one daughter Jane DeWitt Randel and referring to her never as plain Jane but always as Jane DeWitt. Randel later honored one of his sons by naming him Richard Varick, after DeWitt's son. However it came about, DeWitt "had this very personal relationship with John Randel," says Jo Margaret Mano, who became familiar with Randel when she was developing a bibliography of maps in the New York State Archives, because he kept appearing in DeWitt's records and correspondence. "I believe he really looked at Randel as a kind of son, a really trusted individual." In one of his letters to Randel, DeWitt wrote informally of his travels and of the weather—with surveyors, always the weather. Little of business appeared in the document—no instruction, just an update, which does suggest a friendship behind the working relationship:

> As the mail goes from this to Albany only once a Week I have not had an opportunity of writing since my arrival here. On the Day we left Albany about an hour before sunset we were stopped by a very heavy Thunder Shower at Duanesburg which left about two inches of snow on the ground the next morning. The second day we went to Cherry Valley and the following night commenced a terrible snow storm which detained us there . . . the day. On the next we started again and the storm was renewed with equal violence. From Cherry Valley to Coopers Town [*sic*] we had above a foot deep of snow. Lower down the Susquehanna the snow had not been so deep & the day after disappeared on

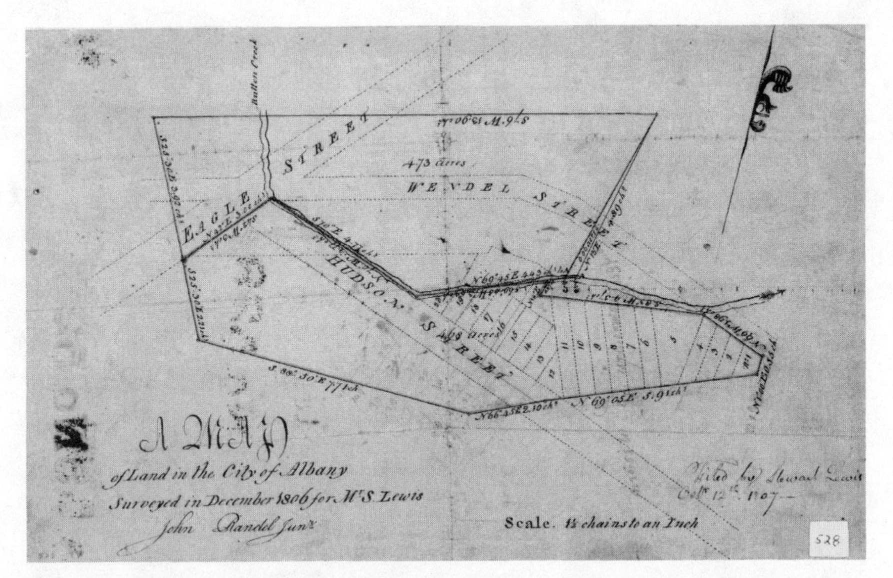

Map of land surveyed for Mr. S. Lewis in 1806 by John Randel Jr. Courtesy of the Bureau of Land Management, New York State Office of General Services in Albany.

our journey. The road from Coopers Town to Owego (excepting from Ogliquoyo to Chenango) were excellent."

In another letter, DeWitt asked Randel to "give my love to the boys and tell them not to be out too much in the streets but study their books. DeWitt Bloodgood and Abraham Hutton both spoke pieces for us this afternoon and did it very handsomely." Randel felt fondness for DeWitt as well. "I have always been in the habit of looking up to you," he wrote in a draft letter.* "It is my pride to have it known that I was brought up under your eye."[36]

The earliest record of Randel working in DeWitt's office is in

*Very few of the letters Randel sent, and none that he received, have come to light yet. But several dozen of his draft letters survive in his field books. It is not often possible to verify whether those letters were sent—or substantially altered if they were. Such drafts nevertheless remain significant and interesting because they reveal Randel's thought processes and serve as a kind of journal entry rather than an official document.

March 1804, the same year captains Meriwether Lewis and William Clark set out to find the Northwest Passage, the year after America had acquired some 800,000 square miles in the Louisiana Purchase. Randel was sixteen, and if he read the newspapers he would have seen reports about the "party of discovery" and a dispatch about great falls, rivers, cliffs, buffalo, and Indians. ("The higher up they went, the more friendly they found the savages, and the better armed," reported the *National Intelligencer and Washington Advertiser* in 1805.) Randel's initial duties were cartographical. Using notes and sketches from surveyors in the field, he made maps of land in the Adirondack Mountains: Scaroon Tract and Brant Lake Tract. He drew a map of the northwestern portion of the Oneida Reservation, based on field reports collected by Charles C. Brodhead. In 1805 he filed a map of Cosby's Manor in Albany and surveyed the Albany-Schenectady Turnpike and the Great Western Turnpike, which ran from Albany to Cherry Valley and Cooperstown. In 1806 he surveyed land in Albany for a Mr. S. Lewis. By 1807 Randel regularly reported to DeWitt from the field with Brodhead as they gridded the central part of the state. In 1808 he mapped an Albany estate for the heirs of Philip Schuyler, an officer in his father's regiment and a former U.S. senator.[37]

A newspaper article published when Randel was twenty-six pushes his start date even earlier than 1804. In that *New-York Evening Post* article, Randel, incensed by a rival's intimation of inexperience, wrote, "I was not hurried into practice after studying only ONE year, or PART OF a year. I served nine years to my profession, of which more than six were under the Surveyor General, since which I have practiced six years." If his account is correct, he started working as an assistant in the field at the age of twelve, soon attracting DeWitt's attention or the attention of someone who recommended him to DeWitt. His start at twelve may explain why college registries have no record of him.[38]

Field notes indicate that Randel was frequently partnered with

Brodhead, who likely served as a mentor as well, having been in the field at least ten years longer than Randel. Brodhead, the son-in-law of land magnate Johannes Hardenbergh, was known to be keenly intelligent. He later worked on the Erie Canal—as did, it would appear, almost every able-minded New York surveyor of that era. What Brodhead thought of Randel is not known; his surviving letters do not say. For his part, Randel seemed warm, or at least courteous and solicitous, toward Brodhead, writing: "I am much pleased to hear that your health is so far restored; please accept my best wishes for your health and happiness."[39]

Randel, Brodhead, and their colleagues followed DeWitt's requirements, and the small army employed by the surveyor general visited and mapped thousands of lots in New York towns. They produced rough maps, sketches of trees and watercourses. At that time there were few or no prior maps, no records to turn to. (Colonial-era maps were printed in Europe and often inaccurate.) Native Americans did not map in the way Europeans did. Their maps were sometimes gestural, sometimes oral, and when pictographic—drawn or etched on bark, hide, or dirt—concerned natural features, events, the spirit world and the afterlife, communities, and resources instead of ownership and property lines. Some of these maps were doubtless communicated to members of DeWitt's office, particularly because his cousin, Moses DeWitt, worked closely with Six Nation tribes and was, according to many accounts, highly respected by them. But by and large, surveyors had to experience firsthand the characteristics of the land by walking it or, at minimum, walking the periphery of the lot they were apportioning.

Nature was noted only insofar as its economic relevance went, as was standard practice for the time. DeWitt required his surveyors to record the land's potential for agriculture and industry and to assign a value based on those assessments, writing that "all Streams Ponds Lakes Marshes Ascents Descents Quality of Soil and Species of timber mostly growing on it; All improvements houses roads local advantages and generally every thing of note must be entered in the field books . . . The Contents of Fields must be given in acres and Decimals and not as formerly in Acres Rods & perches. All circumstances such as the quality of soil advantages of situation Millseats Mines or Minerals Timber & must be considered in making valuations." Because of their important legal role in land claims, surveyors took oaths of office, just as politicians, judges, and doctors did. DeWitt's deputies were sworn in, as were Randel's assistants, some simply signing X, either in his field books or on formal legal documents such as one in the New-York Historical Society collection: "We the Subscribers do Severally Swear that We will faithfully execute the trust reposed in us as Assistants in the Surveys to be performed on New York Island by John Randel Jun. So help us God. Sworn this twenty fourth day of April 1812 before me Jeremiah S. Drake, Master in Chancery." (The master in chancery officiated at a court established by the British that dealt with lawsuits and cases entailing damages. New York State disbanded the court in 1846.)[40]

The description of lot number one in the Oneida Creek Tract, which contained 153.4 acres, was typical of the field book entries: "The westerly part of this lot is well watered by the Crooked Creek—the lot is considerably elevated except in the northeasterly part where there is some low ground. Timber beech, birch, bass, maple, hemlock, & There is about 2 acre cleared and a log house built by Matthias Morris in the southeasterly corner of this Lot." Although the description was by Randel, it could have been by any of his colleagues. The measurement of a lot was basic. The surveyors started in one corner and then, using a compass to find their bearings, measured the length of one side, using a chain. DeWitt's instructions did not mention a surveyor's cross, but crews may have used one to get the all-important right angles. The device could have been quite simple: two small boards nailed together in the center at right angles to each other, with a sight vane attached to (or carved into) the end of each of the four arms.

Sights were very narrow thin slits, confining the field of vision to a specific target. The surveyors would set a line in one direction and then sight a line at 90 degrees to the initial line. Other instruments could also aid in setting a right angle.[41]

As they measured lines, surveyors narrated their tour of the periphery: "Beginning at the northwesterly corner of the tract, a willow tree on the southerly shore of the Oneida Lake marked CCB 1807, Oneida R., JR 1811, and running thence along the westerly bounds of the tract south 32 degrees, east 22 chains and 38 links to a beech tree marked 1.10 on the northerly side of the State Road so called. Thence along the northerly side of said road eastwardly to a stake 31 links west of a beech tree marked 1.2.10.11 thence north 47 chains and 11 links to the Oneida Lake a stake 55 links north four degrees east of a white oak tree marked 1.2 and thence westerly along the shore of said lake to the place of beginning."[42]

Such narratives are the mainstay of early surveying documents, whether metes and bounds or cardinal-rooted grid. They exist in the thousands in state offices throughout the country and contain extraordinary historical data as well as the possibility of wonderful imaginings. Many such descriptions can be laid down on top of today's houses, streets, and strip malls, restoring the eighteenth- and nineteenth-century landscape. These narratives are like the Land of Green Ginger, described in Noel Langley's 1937 children's book of the same name, which floats and then settles atop cities or deserts, transforming them for a time into lush forests. They offer the possibility of seeing the landscape afresh.

Randel's upstate maps, housed in the New York State Archives, in the New York State Bureau of Land Management in the Office of General Services, and in the Onondaga Historical Association, are vivid documents. Fences stand upright, seeming three-dimensional; trout brooks and creeks roam across pages; cross-hatching raises hills. The remote life of farmers and early settlers comes alive in the portrayal of towns comprising just a handful of houses, a place of worship, and long, unadorned roads leading out and in. Whether of Lot No. 94 in Homer, of the villages of East and West Oswego, of the Onondaga Salt Springs Reservation, or of the village of Oneida Castleton, the maps communicate elegance and precision. Clarity and exactness defined Randel's surveying and cartography early on, and they are likely what drew DeWitt to Randel and qualified Randel, in DeWitt's mind, for the New York City project. For DeWitt seemed to have found in Randel what Erskine had found in him: rationally driven exactitude, the desire to improve upon what they had, what they encountered. It was the aspiration, the organizing principle of their time. DeWitt sent Randel to organize Manhattan thus.

The Evil of Confused Streets

"We have suffered so much from pestilence, We have so severely felt the evil of confused Streets, that we have considered the widening of our narrow passages, and the formation of open places and squares, as of the first magnitude and importance. While the eye is gratified, the air is enabled freely to circulate and by its Salubrious operations the City is Cleansed and ventilated." A group of New Yorkers wrote this in a letter to the mayor and the members of the Common Council not long after the 1811 grid plan was unveiled, and it captured some of the impulses that gave rise to the plan in the first place. New York City was growing rapidly in the early nineteenth century; "Few cities in the world present so desirable a scite; few for which nature has done so much," the same letter noted. The city's population had doubled between 1790 and 1800 to 60,000; by 1865 it would reach one million. New York was a busy and, by many accounts, chaotic place. Dogs and hogs roved the streets, scavenging and sometimes attacking; the Common Council often approved expenditures for killing dogs and controlling hogs. Excrement overflowed from privies. Stagnant water stank and sustained mosquitoes. Yellow fever and other diseases—cholera, consumption, typhus, smallpox—swept the city, killing thousands. Mounting refuse and offal could not be controlled despite the hogs and the appointment of a city scavenger. Sidewalks were unevenly kept. Some streets were paved, some not. Fires were frequent (137 between 1810 and 1814, by one account) and could readily leap across the small streets.[43]

City government needed to address all these issues as well as the challenges of poverty, crime, commercial growth, and dock building. The city had lost the nation's capital to Philadelphia in 1790, the state capital to Albany in 1797, and by many historians' reckoning was coming to define itself as the commercial capital. Its port was growing rapidly; within a decade it would handle more than one-third of the country's imports. To encourage commerce and to take on expanded civic duties, however, the government needed revenue.[44]

In 1794 the Common Council hired surveyor Casimir Goerck to map the city-owned lands for intended sale and lease—something he had done once before as well. Goerck's 1796 "Map of Common Lands" would be familiar to contemporary New Yorkers. Seventh, Sixth, Fifth, Fourth, and Third Avenues ran the length of the island, forming a grid. The streets ran perpendicular to the avenues and from river to river; the cross streets were 60 feet wide and 200 feet apart. Several years later, members of the Common Council noted that getting accurate measurements had been almost impossible, as "the Surveys made by Mr. Goerck upon the Commons were effected through thickets and swamps, and over rocks and hills." Those physical challenges had paled in comparison to political ones. Bringing private property into city-mandated alignment was far from popular, according to scholars, including Edward K. Spanne. Some landowners were intent on dividing their own land with streets of their own design. Others worried they would lose boundaries or buildings

Ferdinand Hassler. The National Oceanographic and Atmospheric Administration.

if the city regularized streets and lots.[45]

In 1806 the Common Council passed an act to plan the development of private and public land, commissioning Ferdinand Hassler to do a "proper" survey of the island. But Hassler fell ill "the very day he was to set off," reported the acquaintance who had secured the commission and who found Hassler to be "one of the most interesting foreigners we have for a long time had amongst us." The Common Council would not wait for this interesting foreigner. Outcries by landowners raised pressure on the city government to act. And the struggles were not just with landowners but between aldermen themselves. Unless a higher authority was invoked, the Common Council recognized, a new set of elected members might overturn any plan previously approved. The aldermen and the mayor turned to the state to appoint a panel of street commissioners to devise a plan for the city that would "unite regularity and order with public convenience." It ensured some control by recommending the appointment of three particular "fit and proper persons."[46]

The state legislature passed "an Act relative to Improvements, touching the laying out of Streets and Roads in the City of New-York, and for other purposes" on April 3, 1807. The act specified that the streets and "great avenues" were to be not less than 60 feet wide, as Goerck's grid had been. It decreed that the new plan was to start irregularly—that is, north of a line running inland east from the Hudson River at what is now roughly Gansevoort Street, then south along Bowery Road (now Fourth Avenue), and then east along North Street to the East River. In other words, earlier city development was not going to be upset by or unified with the new grid.

The act followed the Common Council's recommendations for the "Commissioners for Laying Out Streets and Highways," appointing influential, politically savvy men who excelled in different spheres but who in aggregate had the requisite skills and political clout to ensure that the grid was adopted no matter what opposition arose. John Rutherford was a former U.S. senator and New Jersey businessman who had his eye on canals and internal improvements in that state. Gouverneur Morris was a New York City resident, former senator, and drafter of the federal constitution who had extensive holdings in what is now the Bronx (the site of his 1,900-acre estate, Morrisania) and in upstate New York. Morris must have especially deplored the island's remaining wild patches, for he saw untamed landscapes as a force that could "unhinge the intellect," as historian Richard W. Judd writes in *The Untilled Garden*. Judd quotes an 1806 pamphlet by Morris to that effect: "Those awful forests which have shaded through untold ages a boundless extent . . . dazzled the eye of reason and led the judgment astray." DeWitt was the third man chosen. He had the surveying expertise and experience in planning communities in upstate New York. And, ultimately, DeWitt had the perfect man for the Manhattan job.[47]

Randel's work on the island can be divided into three stages, characterized by increasing precision and obsession and by growing public and personal turmoil. The first period ran from 1808 to the end of 1810. During that time Randel measured along roughly east-west and north-south lines, getting widths and lengths for streets and blocks using basic equipment. The streets were to run river to river to facilitate traffic to the docks, and apparently to ensure cross breezes to rid the city of stagnant, unhealthful air.

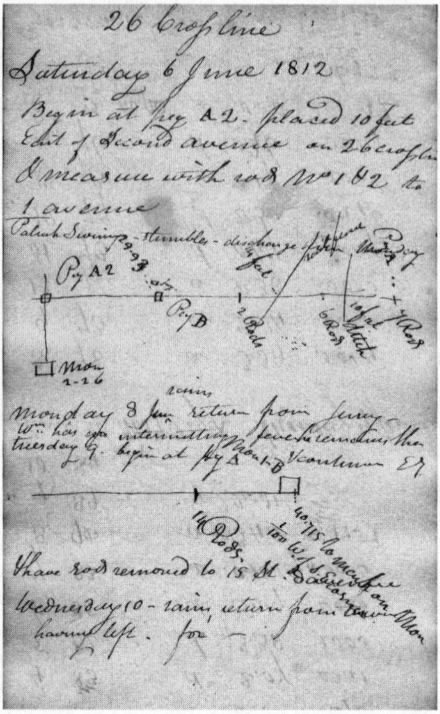

(The main axes, or avenues, run nearly 29 degrees east of true north, and the streets, rotated the same amount, run nearly 29 degrees north of true west. The avenues and streets therefore fit the tilt of Manhattan itself.) Using those measurements, Randel produced three copies of a manuscript or unengraved map. This was the famous Commissioners' Plan of 1811. The second stage ran from early 1811 to about 1817, when Randel designed and used more sophisticated instruments and equipment, and during which time he inscribed, scratched into the dirt and rock of the island, the 1811 plan. Randel conducted a geodetic survey—sophisticated, accurate, recording elevation, not just distance—as opposed to a regular land survey. During the last period, which ran from 1818 or so to 1821, Randel, aided by his first wife, became a cartographer of the highest quality.

Page from John Randel Jr. field book. Courtesy of the collection of Mark D. Tomasko.

Much of what we know about Randel's Manhattan surveys comes from two principle sources. The *Minutes of the Common Council of the City of New York* contain records of Randel's reports to the committees on surveys and on roads and streets; most of those records concern finances, contracts, and deadlines. The more complete source of information resides in Randel's field books, housed at the New-York Historical Society and donated by Joseph O. B. Webster, a former city surveyor and engineer. According to *The Iconography of Manhattan Island*, by Isaac Newton Phelps-Stokes,

Randel's second wife valued the notebooks at $2,500 and put them up for sale around 1870, five years after Randel's death. They had not sold by 1877, when she again approached the Board of Aldermen to sell them, this time for $5,000. Perhaps the board did buy them, and perhaps that is how the field books ended up in a city employee's care. However it happened, Webster turned the valuable city records over to the historical society in 1913, trunk, insurance policy, accompanying maps, and all. Usually field books for official surveys remain the property of the government commissioning the work, and it would be unusual for an agency to pay for them. Randel clearly considered them his property, however, and they traveled with him from home to home. One of the New York field books even contains notes from 1861, which Randel wrote while in Delaware. His attachment to his field books—possibly a shrewd professional strategy—proved nearly disastrous for New York State at one point in the 1830s.[48]

Almost all forty-five field books are thick, brown, and leather-bound. Each opens like a reporter's notebook; Randel ordered them from Obadiah Van Benthuysen, an Albany printer. They are simultaneously sparse and rich documents. In some places they are professionalism embodied: figures, sine and cosine tables, measurements, sketches and diagrams, a log of ground covered and markers set. Others contain those careful records as well as tantalizing insights into Randel's personal life between 1808 and 1823, because he drafted letters, copied contracts, mentioned family and friends, and recorded anecdotes, travels for work and pleasure, and recipes.

Randel's field books are above all a personal palimpsest of visiting and revisiting, recording, checking, correcting, certifying. It is not unusual for one book to contain entries from as many as four years layered over each other. Irregularities unsettled Randel. His field books are filled with corrections and adjustments and recertifications of his corrections and notes about why measurements were off. Every field book attests to his precision and pursuit of perfection: in his work, his expenses, his schedule. Even a rumination on one of his men's errands offers a window into Randel's exactitude—and, it seems, his sense of humor:

> Clarkson left his gun in Bowery, 1,000 feet from hence. Fearing he will not have time to get it in season to parade, requests the loan of my horse and saddle, at 6:40 A.M. At 7:07 starts and 8:12 returns. He is going and returning 65 minutes, and 32 minutes in getting and putting up the horse. The whole distance he could have walked in 43 minutes. He with the aid of the horse employs 97 minutes. He is therefore 44 minutes longer going with the horse than he would have been without. By loaning the horse I have therefore damaged him 44 minutes of time, which I will allow him for. Conclusion: it is bad policy to let my horse but for my own convenience.[49]

With Hands, Feet, Swords, Staves, Sticks, Stones, Knives and Axes

The street-plan legislation passed in April 1807, but little seems to have been accomplished until the following year, when Randel was called in. The slow start may have been partly due to the scorching heat that first summer—perhaps what the commissioners meant by "reasons particularly unfavorable" when they described the delay. Incompetence was another problem. In October 1807, Morris complained to DeWitt about one of the city surveyors, either Adolphus or Charles Loss. According to Morris, Loss made a very basic mistake—taking magnetic north to be true north—and when Morris had him compare different surveyors' compass readings with old city maps, Loss discovered (not to Morris's surprise) that his maps were incorrect: "It turned out again (as I apprehended) that the old Maps are erroneous and that his work from them must be done over." Morris also noted that two of DeWitt's upstate surveyors, Cockburn (either James or William Jr.) and Beach, were busy laying the island's meridian, or north-south line, "which is you know a tedious Business." Morris's tone was urgent; he appeared upset about Loss's lack of accuracy and its consequences: "We cannot indeed remedy What is past, but Mischief may be prevented in future. Will you my dear Sir pardon me for suggesting that an official Representation from the Surveyor General might not be improper."[50]

DeWitt apparently did go visit Loss. Loss seems to have been sacked, and the commissioners then passed over other surveyors who knew the city well, including William Bridges, who was to become Randel's rival. Instead the commissioners brought Randel down from Albany as "Secretary and Surveyor to the 'Commissioners of Streets and Roads . . .' and also as their Chief Engineer"—Randel's own description of his title. The first record of his work on the island appears in July 1808.[51]

Randel's initial surveys resembled his upstate jobs in that he established straight lines and used standard equipment to do so. Only three of the surviving field books of the Manhattan work record data from fieldwork between 1808 and the end of 1810. A typical entry did not contain great detail; one read, quite simply, "Exploring a line run on a course (S41W) parallel to the streets at Kipps Bay—from North Street to Bellevue." Nevertheless, it is clear that Randel measured crosslines (streets) running from the Hudson River to the East River at right angles to avenues. He must have set the orientation of the avenues and the streets with a compass, having established with the commissioners, or having been told by them, that the avenues would follow the island's easterly skew. Randel would walk a line, as he referred to a compass, measure that line, note and sketch what he encountered. As he walked, he and his crew pushed marker pegs into the ground at regular intervals so they could keep track of the line and record distances.

Randel made his measurements with chains and with 10-foot rods and those of other lengths. He used a transit telescope, sometimes simply called a transit, which can flip in the vertical plane or rotate in the horizontal, so a surveyor can look at one point along the line and then turn or transit the telescope 180 degrees to look at another point. By repeatedly foresighting and backsighting, as this process is called, surveyors could set a straight line. Randel's use of the transit would have been de rigueur for the era.[52]

When Randel established right angles, he must have used his compass, a surveyor's cross, or a theodolite. A theodolite comes in many different forms but typically combines a compass and one or two sets of sights, which could be set at right angles to one another, and sometimes a telescope. Surveyors use theodolites for recording horizontal and vertical angles. Randel's crew included chain men, ax men to clear the path, and flag men. Flag men would semaphore between the point of origin, where the instrument (telescope, theodolite) stood, and a distant point in the direction of the survey. Randel noted the signaling system: "Up—wave to the right of Instrument. Down—ditto . . . left . . . ditto. Right—wave the flag to right and left when the flag at the instrument is moved up and down continuously." The telescope would fix or sight on a marker—it could be a flag or a pole—in the distance along the survey line. Until that marker was set in exactly the right spot, the flag men would communicate about how it should be moved.[53]

Theodolite, circa 1800. Courtesy of Jeffrey Lock, Colonial Instruments.

One account says Randel "carried ink bottles suspended from his neck so that his hands might be free to take notes, or perhaps hold the chain or rod," but the source of that twentieth-century newspaper account is not cited. Having free hands certainly would have been useful. One line took him through, over, or around a barracks, a powder magazine, asparagus, carrots, clover, potatoes, fences, turnips, grass, a swamp, a stone wall, and corn. He passed privies and ponds. He did not plumb all ponds. Of one he noted, "Water pretty deep. It is said." Although he occasionally noted a high watermark and a low watermark, he did not record elevations. His early descriptions included "gentle ascent," but no precise height, which was consistent with his training and with the approach that most surveyors followed. Randel marked rocks and trees—in one case he drew an apple tree lying flat with a cross mark on it—and sometimes chopped down trees, which caused an uproar. Randel recorded landowners' names and sketched the structures on their land and the distance of their houses and stables to existing roads and pending roads.[54]

Randel worked nearly every day. He stayed in the city, perhaps walking through narrow streets poorly lit by whale-oil lamps when he returned late from the field. His headquarters were the commissioners' office, "*then* in the country," on the northeastern corner of Christopher and Herring (now Bleecker) Sreets. The office was a room in the house of Amasa Woodsworth, "a very worthy mechanic," and his family. (The office moved in 1809 to a room in Smith's Tavern on the Eastern Post Road, at roughly Third Avenue and 77th Street.) Randel's eye for detail was not limited to landscapes and distances. "I almost daily passed the house in Herring Street where Thomas Paine resided," he wrote, "and frequently in fair weather saw him sitting at the south window of the first story room of that house. The sash was raised, and a small table or stand was placed before him with an open book placed upon it which he appeared to be reading. He had his spectacles on, his left elbow rested on the table, and his chin rested between the thumb and fingers of his hand; his right hand lay

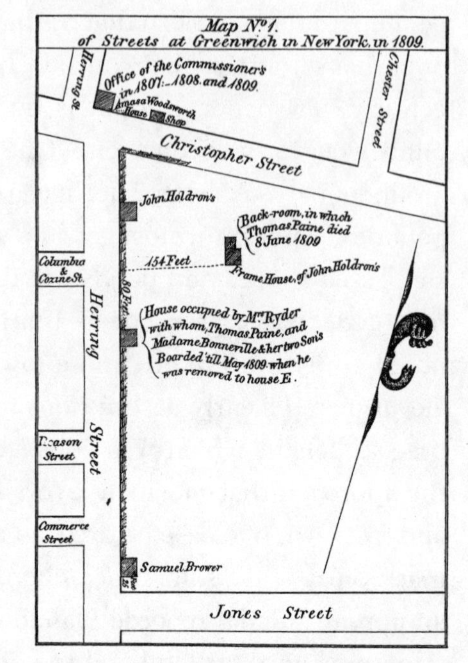

Maps of Christopher Street in 1809, 1827, and 1864 by John Randel Jr. From Maud W. Goodwin, Alice C. Royce, and Ruth Putnam, *Historic New York*.

upon his book, and a decanter containing liquor, of the color of rum or brandy, was standing next his book or beyond it. I never saw Thomas Paine at any other place or in any other position."[55]

Randel and his host, Woodsworth, differed vehemently on Paine's *The Age of Reason,* as Randel later wrote, "he advocating and I opposing it." Woodsworth was not alone. For a time, thousands of Americans responded to Paine's plainspoken arguments against religious institutions and the Bible, and *The Age of Reason* appeared in seventeen editions. For his part, Randel loved reason deeply. He was a man of science. Yet he felt his faith just as deeply and was having nothing to do with Paine's deistic thesis, noting,

> One day in June, 1809, after I came from the city to our office, I stepped into Mr. Woodsworth's shop, when he informed me that THOMAS PAINE *had just died*, and that he had been with him the previous night; and I think he said *he had just* come from there. I inquired of Mr. Woodsworth, whether Thomas Paine continued in his belief of the doctrines advocated in his "*Age of Reason*;" he replied, that Thomas Paine did not recant, and that he (Paine) declined speaking upon that subject. I informed him of the rumor that Thomas Paine's friends

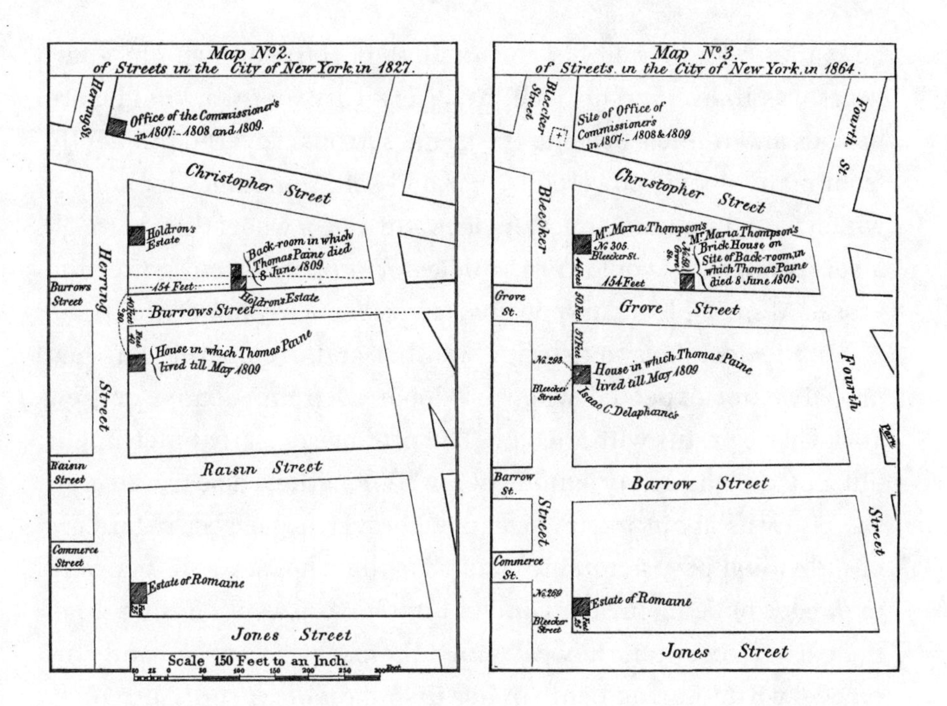

and disciples, who had charge of him during his last sickness, had refused admittance to some clergymen and others who desired to see and converse with him; he replied, that some meddlesome persons, taking advantage of his supposed debility of mind, as well as of body, called to see him, and they were very properly refused admittance. I also mentioned the report that Thomas Paine's friends had kept him partially under the influence of intoxicating liquor, to prevent him from recanting, or making any confession of his error, or regret, for having published his "*Age of Reason*." He said there was no truth in such report, and that Thomas Paine had declined saying anything upon that subject."[56]

Randel assiduously attended church on Sundays and respected the Sabbath. He often took a ferry to New Jersey, where his mother's family lived and where several relatives were elders in the First Presbyterian Church in Orange and Bloomfield. During these visits he came to know well his cousin Matilda Harrison, who in a few years was to become his wife. New York City's ferries, vital before bridges and tunnels became the island's axons, voyaged frequently from many docks, powered by oar and sail, sometimes by horses, which would walk in circles on deck, turning a wheel that, through a series of gears, would turn paddles or propellers and power the vessel. After 1811, steam-powered ferries regularly joined the array.

Randel lost days to extreme weather and to something he had probably not expected when he departed from Albany for New York City: run-ins with the law. The promise of a great metropolis did not delight many landowners. "When they discovered that the city was about to run streets wherever it pleased, regardless of individual proprietorship, and that their houses and lots were in danger of being invaded and cut in two, or swept off the face of the earth altogether, they esteemed themselves wronged and outraged," wrote Martha Lamb in her *History of the City of New York*. "At the approach of engineers, with their measuring instruments, maps, and chain-bearers, dogs were brought into service, and whole families sometimes united in driving them out of their lots, as if they were common vagrants. On one occasion, while drawing the line of an avenue directly through the kitchen of an estimable old woman, who had sold vegetables for a living upwards of twenty years, they were pelted with cabbages and artichokes until they were compelled to retreat in the exact reverse of good order." Randel was frequently hauled off, and a former mayor had to post bail for him when the commissioners were not in town. As he noted, "I was arrested by the Sheriff, on numerous suits instituted against me as agent of the Commissioners, for trespass and damage committed by my workmen, in passing over grounds, cutting off branches of trees, &c, to make surveys under instruction from the Commissioners."[57]

Court documents captured some of the landowners' passion and anger. In 1808, for instance, John Mills brought a case against Randel for damage to his farm in the city's eighth ward, now the West Village. Randel's field book for August recorded nothing unusual, whereas for Mills, August 26 was an apocalypse. The standard language for a trespass suit of that time was hyperbolic, and Randel's powers of destruction were described as those of a wrathful deity. With "force and arms to wit with hands, feet, swords, staves, sticks, stones, knives and axes," Randel "with his feet by walking trod down and consumed and then and there felled cut down and otherwise injured and destroyed five hundred Ash Trees. Five hundred yew Trees five hundred Elm Trees, five hundred Apricot Trees five hundred Peach Trees," and the same amount of plum, cherry, nectarine, apple, and pear trees and gooseberry bushes. Randel also "then and there felled cut down pulled up trampled down and otherwise injured and destroyed five hundred Cabbages, five thousand Beetes, five thousand Potatoe Hills, five thousand Carrots, five hundred Pinks, twenty thousand Strawberry bushes and plants five hundred Tulips then and there standing growing and being to the value of five thousand dollars." Through his lawyer, Francis B. Winthrop Jr., Randel pleaded not guilty. Mills, who seemed in his own estimation capable of feeding the entire island and perhaps residents of New Jersey as well, ultimately won. On June 25, 1810, the Supreme Court found Randel guilty and awarded Mills $109.63—somewhat shy of the $5,000 Mills had hoped for, but it covered his legal expenses.[58]

Although they lost, the suit ultimately benefited Randel and the commissioners. The legislature had given the commissioners four years to finish the plan. Surveying was seasonal and time-consuming, even without constant delays caused by obstreperous landowners and a sheriff sensitive to their point of view. The Mills case and others like it helped convince the state legislature that the situation was dire, fiscally as well as practically. The commissioners threatened their resignation over the "vexatious interruptions" if Randel and his crew were not protected from arrest, suits, and violence. On March 24, 1809, state legislation permitted the commissioners and those working for them to enter private land during "the day" for the purposes of surveying and to cut trees and do other damage. Owners could bill for damages within a month. "*And be it further enacted*, That if any person shall be sued for any thing done in pursuance of this act, it shall be lawful for such person to plead the general issue, and to give this act and the special matter of defence in evidence, under such plea."[59]

In September 1810 Randel detoured from his New York City work to survey the Post Road between Albany and New York. He started at Albany City Hall on State Street, ferried across Hudson's River to what is now the city of Rensselaer, and proceeded south. He finished at New York's City Hall, recording a distance of 154 miles. Randel suggested changes to the road to reduce mileage. "This is an important fact, and highly interesting to the citizens of Albany and New-York, as it entitles them to a reduction of one-third of the present rates of postage between the two cities," commented the *Albany Gazette*. (Alterations were later made, and mileage and postage fell.) Also interesting is the fact that the 23-foot-long map may well have been the first draft of the final New York City grid plan. Randel showed the post road entering a gridded island. The road traveled in faint dashed lines underneath a planned city.[60]

Although Randel may have completed this initial draft in early fall 1810, several more months of work on the official version remained. There was still no final map when in November Gouverneur Morris reported to the Common Council that the commissioners had completed their duties. The streets and avenues had been measured and the future city had form as far north as 155th Street, not beyond. "To some it may be a matter of surprise that the whole island has not been laid out as a city. To others it may be a subject of merriment that the Commissioners have provided space for a greater population than is collected at any spot on this side of China," wrote the commissioners in a published report. Whether perceived as too great or too small, the plan remained unfinished. As Morris noted, "Much is yet to be done on the ground." Before the grid could be physically anchored to the island, the city government needed to set permanent monuments to delineate and fix the streets and avenues so they could be constructed, graded, and perhaps paved. Morris said that some 100 miles needed to be measured with extreme care so that 3,500 monumental stones could be set at future intersections.[61]

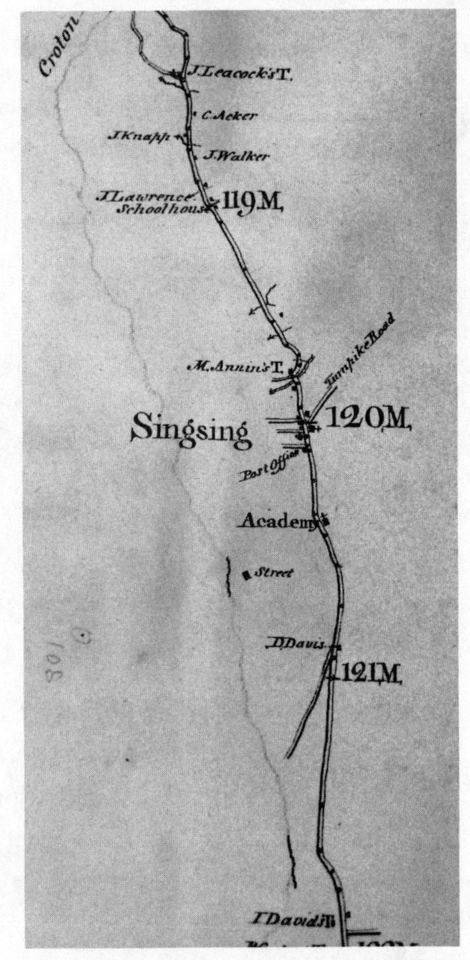

Detail of Singsing, New York, on the 1811 Albany Post Road map by John Randel Jr. Courtesy of the Albany County Hall of Records Archival Collection.

The commissioners recommended to the council a sole surveyor who could obtain the accuracy "requisite in a work of this sort, where the difference of an inch may afterwards be a source of contention." That sole surveyor was Randel. The commissioners' endorsement did not prevent city surveyor William Bridges from throwing in a bid as well. Bridges's petition and another he placed just on its heels were noted and brooded on by Randel for several years before his anger flared into public view. The Common Council followed the commissioners' recommendation, passed on Bridges, and on December 31 signed a provisional contract with Randel. A permanent one was promised once he had finished the three copies of the map of the Commissioners' Plan. The aldermen desired the map in hand.[62]

Randel submitted the map three months later, on March 22, 1811. The commissioners affixed red wax seals to the nearly

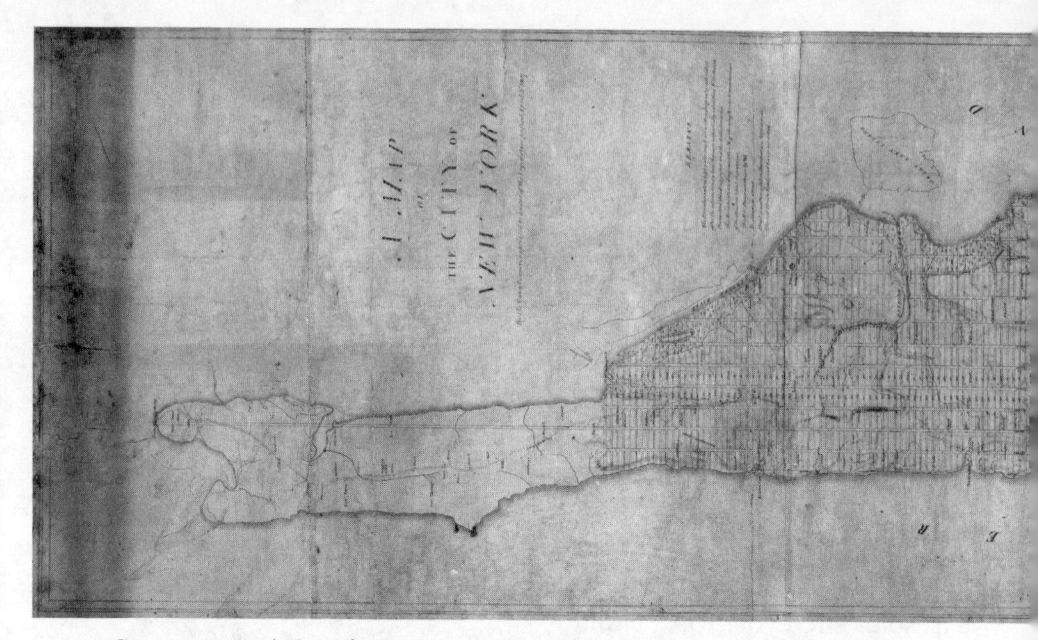

Commissioners' Plan of 1811 by John Randel Jr. Manuscript map from the collection of the New York County Clerk; image courtesy of the New York City Municipal Archives.

9-by-3-foot map, and Archibald Campbell of the surveyor general's office in Albany, New York City's master in chancery, and mayor DeWitt Clinton signed the document. The clerk of the Common Council filed the map on April 1, "which was *two days prior to* the expiration of the Statute," Randel later noted—the italics his own. The deadline had been met, but barely. Today Randel's three copies, somewhat the worse for wear, live in the New York City Public Library, in the municipal archives, and in the office of the Bureau of Land Management in Albany. The streets-to-be are inked; the streets-that-were are faint lines meandering tentatively here and there under the confident grid. Hundreds of tiny boxes—homes, churches, taverns—are scattered around, some sitting inside blocks, some in the middle of an impending road, all exhibiting an unaligned, insubstantial quality. The island north of 155th Street looks naked, smooth, like a thin clean bone jutting out of a plump drumstick.[63]

In June, a few months after the map had been delivered and presumably studied with care, the council assembled a general

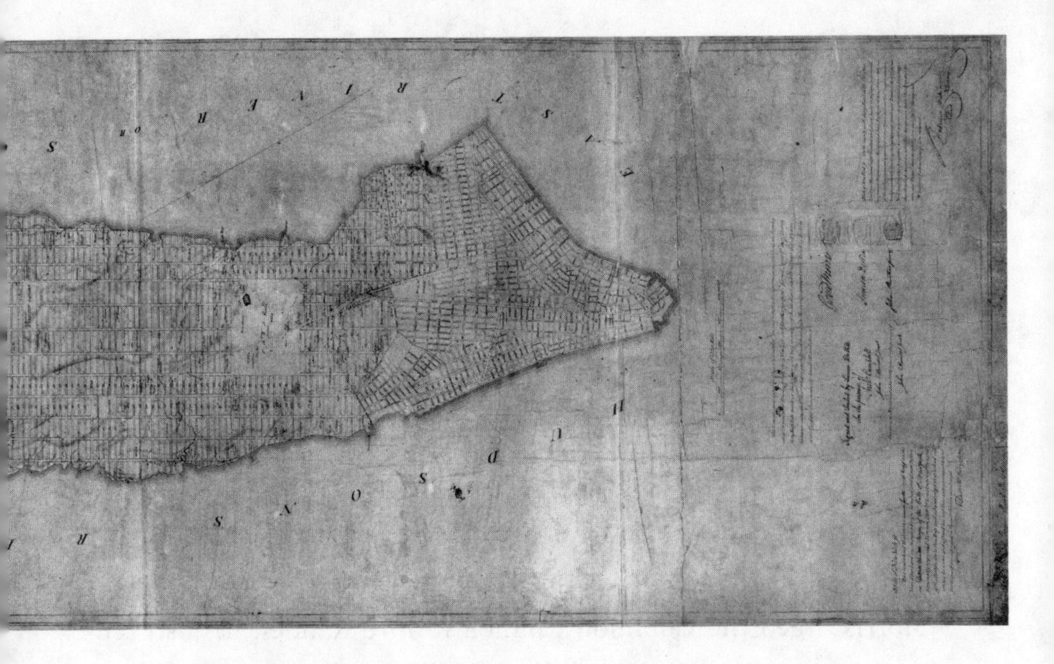

superintending committee of five aldermen to oversee Randel's work—among them Nicholas Fish, a lawyer and one of Randel Sr.'s commanding officers. Randel was twenty-three, still a young man, one whose professional reputation was glowing, ascendant. He had, apparently gracefully, weathered lawsuits, arrests, and angry New Yorkers. He had won the trust of his mentor's powerful friends and colleagues. He had completed his fieldwork on time and had drafted three copies of a beautiful, masterful map. His employers viewed his accuracy as unimpeachable and key to the next stage of city planning.

That June, as all involved contemplated the task ahead, neither Randel nor the members of the Common Council could have known how vastly time-consuming, complicated, contentious, and costly Randel's next job on the island was to be. As massive an undertaking as it had been to survey the island to make a blueprint for future development, it would not compare to the challenges of realizing that plan on the ground and marking lines on earth instead of paper. The stunning map Randel had created, the famous and infamous 1811 map that gave rise to New York's geometry, was fantasy. The island was hilly and stony, woven with creeks, soft in places with beaches, marshes, and wetlands. For the grid to come to life, it had to disrupt and disturb. It had to blow apart rock, tear up earth. It had to scar.

Upon the Discovery of This Strange Occurrence

During his first two years on the island, Randel realized his equipment was not up to the Manhattan terrain, and so when Morris urged the Common Council to hire Randel, he also recommended that Randel be given $1,000 for instruments (about $17,700 today). At that moment Randel ceased being just a virtuoso surveyor. He became an engineer. Rather than purchasing instruments, he designed his own, and his decision to do so was revealing. With that act Randel asserted his mechanical self-confidence and mastery, applied his engineering genius, and committed to an ever-deepening obsession with perfection. Before those instruments existed, Randel might bemoan the limitations of his craft, but if he had done his part correctly, failings could be attributed to technological imperfection. By creating instruments that could achieve his dreamed-of accuracy, Randel would have had to shoulder all shortcomings; he would have, from this point on, only himself to blame. Obsession and brilliance were—and are and ever will be—double-edged. Randel came to feel the sharp side of the blade.[64]

Randel created seven instruments for his Manhattan work. Some of them resembled contemporary surveying instruments, but they were unlike them in significant ways. Jeffrey Lock, an expert on Colonial surveying equipment, describes Randel's innovations as "monumental." Lock, one of the country's best-known restorers and repairers of early scientific instruments, says he rarely

Transit telescope by Richard Patten, circa 1850. Courtesy of Jeffrey Lock, Colonial Instruments.

engages with post-1795 instruments, because later instruments lack the quality and artistry he so admires in the Colonial ones. But Randel's instruments, which Randel illustrated in the upper right corner of a self-published 1821 map and which he celebrated as his "inventions," are unique and ingenious. "I have never seen this before. These are very, very large serious instruments," Lock says. "This guy was a mathematical, geometric genius."

Before Randel developed his new tools, he had been using, as mentioned, a chain and rods for measuring, as well as a wire measure in some instances, to check the length of the chain and rods against. He may have had a theodolite. He had a compass. He had a transit telescope. He also used one important new innovation, the vernier compass. This compass was an American solution to a problem plaguing all surveyors, but of particular concern in the United States in Randel's lifetime.

All surveyors relied on compasses. The trouble was that magnetic north was and is nomadic, wandering. Today, for instance, the magnetic pole is moving about 40 kilometers a year through northern Canada, bound for Siberia. True north—the North Pole, the tip of the earth's axis—reigns on maps, holding steady and fixed, anchoring depictions of the world, of countries, of states, of landowners' lots. Yet true north and magnetic north rarely meet. The magnetic fields of the earth change from site to site, change over time at a site, have different strengths in different places, and sometimes flip completely, reversing north and south. The magnetic fields are generated by a dynamo, by the chaotic roiling and swirling of the molten liquid iron that surrounds the planet's solid inner core; the motion of the liquid metal generates electric currents, which in turn generate magnetic fields. Because of the roiling and the swirling, magnetic fields wax and wane and weave around. Paleomagnetists can see these meanders inscribed in rock. When rock is formed by upwelling magma, the iron components of that magma align with the magnetic field of the earth at that spot. Then the rock cools, its magnetic orientation no longer mutable. A rock formed later, or in a different location, will have a different magnetic signature. The fields and those rocks tug on a compass needle, distracting it from its attraction to magnetic north. A compass can pull whoever uses it in the wrong direction.[65]

Surveyors measure the difference between the geographic North Pole and magnetic north as an angle. The point of the angle, or vertex, is centered in the compass, with one line running from that point to true north, the other line running to magnetic north. The degree of the angle is called magnetic declination or magnetic variation. European explorers and merchants had been aware of magnetic declination since the sixteenth century and compensated by devising tables that mapped known magnetic lines and by using astronomy to determine true north. In one approach, navigators and explorers observed the pole star, or North Star, which aligns almost exactly above the pole throughout the earth's rotations. The difference between the line to the pole star and the line indicated by the compass would yield degrees of declination. Observing the rising and setting sun could also yield true north. Once travelers had calculated true north, they could accurately follow maps and not be pulled off-course by the vagaries of magnetic north.

In England, where property lines had long been marked by hedgerows or fences, magnetic variation did not undermine surveyors. English surveyors accordingly preferred theodolites, using them primarily as protractors to measure angles already established on the landscape by fields, roads, and so on—and only infrequently using the compass in the middle of the instrument.

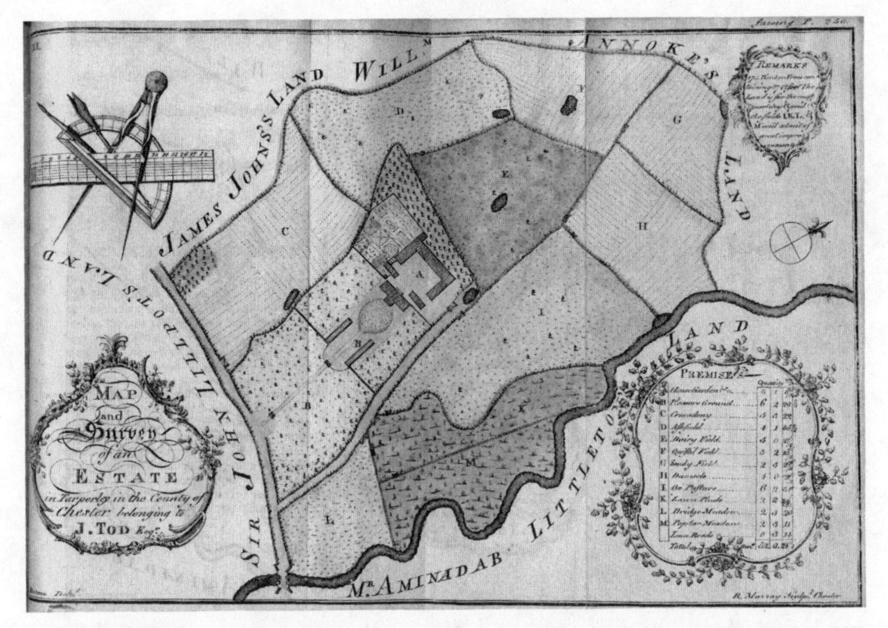

1775 British survey map. Courtesy of Jeffrey Lock, Colonial Instruments.

When Ferdinand Hassler worked in the United States, he had a special carriage constructed for a London-made theodolite, saying that "the inconvenience of transportation and its weight make it very difficult to bring this instrument to the top of our mountains . . . it might weigh without trouble five or six hundred pounds; I can easily move it to any of my stations with my four horses pulling my strong coach upon springs, in which I convey two, three or more instruments, in which I sleep nights and where I now write on a suspended table, which is in fact my house and home and my traveling companion." Hassler's vehicle was mocked by some U.S. government officials for its unusual design and its other cargo: his cheese, crackers, and cuvée.[66]

Few hedgerows, roads, or fences saved American surveyors from the vagaries of magnetic variation. Most surveyors hacking through forest and mucking through swamps preferentially relied on a plain compass because it was hardy and easy to carry. And the compass veered here and there, bound to invisible, shifting

Vernier compass by Lewis Michael, circa 1790. Courtesy of Jeffrey Lock, Colonial Instruments.

magnetic fields. One year the eastern line of a property could run from the barn to the black oak in the back pasture; a decade later, part of that back pasture might belong to the neighbor. Confusion and lawsuits proliferated. In 1772 the Virginia General Assembly responded by passing legislation requiring surveyors to get their surveys right—a futile fiat. The Continental Congress issued a similar requirement in the 1785 land ordinance; the U.S. Congress suspended it a year later, because it was proving impossible to implement. Then, in 1796, Congress reinstituted the requirement because several Americans, including David Rittenhouse, a brilliant instrument maker and surveyor who extended the Mason-Dixon line, had invented a compass that could tackle the challenge: the vernier compass, it was called in New York and other states, and the nonius compass in Philadelphia, where Rittenhouse was among the first to make one.[67]

A vernier is a small additional scale on a measuring instrument that can move and provide finer supplemental measurements. It was devised by Pierre Vernier, a seventeenth-century French engineer. (Philadelphia perhaps felt more kinship to the sixteenth-century Portuguese mathematician Pedro Nuñes, who invented a different gradated device called the nonius.) Combined with a compass, a vernier could be used to get readings to minutes or seconds of degrees. It could also be used to correct for nomadic north. Once true north was astronomically established, surveyors set markers for a meridian they could repeatedly refer to. If the compass needle pointed some degrees to the left or right of the meridian, the vernier scale could be moved to indicate how many degrees to the left or right of magnetic north true north lay. This difference would have to be taken frequently, and the vernier compass gave surveyors that flexibility.

Simeon DeWitt had established a meridian marker in Albany after he moved there in 1785, one Randel was surely familiar with. (Desperate for magnetic information during the Revolutionary War, DeWitt had placed newspaper requests for "ANY MATHEMATICAL GENTLEMEN who can furnish the Subscriber with the correct *variation of the needle*.)" DeWitt routinely checked his compass against Albany's meridian, and devised a rule his surveyors followed: declination varied from the west to the east about 3 minutes of a degree a year—until "to my surprise I found that a sudden change had taken place in the direction of the needle," he wrote of 1807 (meaning, in all probability, 1806 instead). And, he noted, erratic variation persisted. "This irregular daily variation sometimes amounts to a quarter of a degree, and does not appear to be governed by any uniform law, but to be influenced by . . . heat, and probably by other fluctuating causes of which we are ignorant." (DeWitt was right. A contemporary reconstruction of magnetic declination in New York between 1590 and 2000 shows the dramatic shift DeWitt observed. But it was to be well into the next century before scientists knew about the earth's core, the dynamo, and the ever-shifting magnetic fields.)[68]

Randel encountered the irregularity DeWitt had puzzled over as well. He had been out in Albany measuring property when he noticed that measurements he had made two weeks earlier were no longer good. "Upon the discovery of this strange occurrence,

I stopped my work, and returned to the Surveyor General," he later wrote. The two of them examined the compass Randel was using—"it was one of David Rittenhouse's best make," and one DeWitt had used during the war. They found "all its parts were in perfect working order and repair." DeWitt wondered if the total solar eclipse the week earlier, on June 16, 1806, had contributed to the variation. He had commissioned an Albany artist to paint the eclipse and later sent the work to the American Philosophical Society in Philadelphia via Dr. Benjamin Rush. The painting captured, he wrote, "as true a representation of that grand and beautiful phenomenon, as can be artificially expressed. The edge of the moon was strongly illuminated, and had the brilliancy of polished silver. No common colours could express this." During that grand and beautiful phenomenon, DeWitt reasoned, the sun's rays might have dislodged particles from the moon and carried them to the earth, where they dislodged magnetic patterns: "May I be permitted to escape the charge of advancing in absurdity, in suggesting the *possibility* that the lunar effluvia conveyed to the earth by the rays of the sun, on that occasion, might have had an agency in producing the phenomenon I have described."[69]

Although Randel and DeWitt examined the variation every autumn thereafter, they could not easily check changes against DeWitt's previously established meridian. As Albany had grown, the very prosperity DeWitt facilitated had gotten in the way of his old marker. He does not describe the original marker in his writings; it may have been a line marked in pavement or a groove etched into a stone or monument. Whatever form it took, the marker became obscured by buildings, according to DeWitt, and he couldn't get a clear line of sight, which he would have needed to observe the passage of the pole star. Albany's public square, however, apparently provided sufficient view of the night sky. Finally, over four nights in October 1817, Randel and DeWitt used a transit telescope to observe the pole star from the south side of the public square and established a corrected north-south meridian.

Manhattan presented Randel with magnetic challenges as well. The island is the crest of an eroded ancient mountain range, one that rose more than 440 million years ago in what is called the Taconic orogeny. Three kinds of metamorphic rock, deformed and uplifted and eroded across vast periods of mountain building, tectonic-plate shifting, and volcanism that began more than a billion years ago, make up the island. Fordham gneiss is the oldest rock; Inwood dolomite (marble) and Manhattan schist are millions of years younger. The schist gives the island much of its rocky character. The brown or gray mica-flecked rock runs the length of the island, diving deep and out of human range for stretches near 125th Street and Canal Street, surfacing as a substantial hill in Marcus Garvey Park and as big rocks in Riverside Park and Central Park. Just after Randel's time, this particular rocky geology came to support skyscrapers and Manhattan's distinctive skyline. During Randel's time, the rocks' magnetic signatures tugged this way and that as he climbed or circumvented schist formations and large boulders, glacial erratics delivered by the ice sheets.

The tugging and pulling had long been noted in Manhattan. A year after Gouverneur Morris wrote to Simeon DeWitt complaining of Loss's inability to correct for errors of magnetic declination, the Common Council's own street commissioner, John S. Hunn, became increasingly irritated about mistakes in previous city surveys, writing, "The Street Commission has the honor to represent, that the necessity and importance of more correct surveys have become conspicuous to every Builder and person possessing property in this city. That the frequent mistakes which arise are a constant source of continual perplexity to the Common Council, as well as a considerable and unavoidable expense." Hunn attributed the mistakes to missing markers, the "inattention" of surveyors, and, most particularly, to poor or lacking instruments.[70]

Randel was not one of those inattentive surveyors, nor were his instruments poor. Common Council records noted that Randel did not use the ordinary model of compass, which would have led to lines that would "vary in proportion to the distance and variation of the needle owing to the attractive quality of the substances contained in the Rocks & other bodies within its range." He was very much DeWitt's protégé. He must have set a meridian marker in New York City—or used the one DeWitt's office had established in 1807—for he would note how the compass shifted. (Fifth Avenue and Tenth Avenue may have served as his meridians.) His notebooks contain references to readings at different positions as he walked his lines. One August day he wrote: "10 and a half o'clock, needle points S 41.5 W; 12 o'clock S 41.40 W; 1 o'clock need points S 41.40 W; 1 and a half o'clock, S 42.00 W." Another morning at 7:30 he checked his needle against the meridian. "Vernier right," he noted. The vernier solved the problem of ever-changing north and helped Randel obtain and maintain the right direction on magnetically intricate terrain.

Randel's charge was to lay down the city as though it lay on level land, what surveyors call plane surveying or mapping. Randel referred to the practice as "reducing distances to a horizontal measure." It was and remains the standard approach to describing small areas in which the curvature of the earth is not discernible. Over greater distances, surveyors compensate for the earth's curvature by making additional calculations to arrive at a plane, which Randel sometimes did as well. Not only did he have to map to a flat plane and figure accordingly, he had to set perfect 90-degree angles in this hilly terrain, which was challenging. It was his official mandate but also his compulsion. For his own peace of mind or perhaps sense of self he needed to "have this work completed with a degree of accuracy never before attempted, except for measuring a degree of Latitude, or for a Base Line for some nation or other important public purposes." Randel's language echoes Charles Mason and Jeremiah Dixon's; they had measured a degree of latitude as well as established important boundary lines for Maryland, Delaware,

Pennsylvania, and what is now West Virginia. Randel clearly viewed himself as belonging in their eminent company.[71]

For Randel, obtaining that accuracy required not only new instruments but instruments that improved on those then available. Over the course of seventy-four days between late 1810 and the commencement of fieldwork in July 1811, he had seven instruments constructed according to his specifications. No records indicate which instrument maker he commissioned. In later years Randel used a renowned Philadelphia instrument company. It is likely he stayed local for the Manhattan work, and several experts were available. Randel may have contracted the optician Benjamin Pike Sr. or Richard Patten, two well-known instrument makers in New York City with whom he was familiar. (He contacted them both in 1819 about making or purchasing a leveling instrument.) Patten had the later distinction of repairing Ferdinand Hassler's European-made instruments. Whoever made Randel's gear was highly skilled, and expensive. Although given $1,000 by the Common Council, Randel later wrote that the instruments cost between $3,400 and $4,000 (between $60,100 and $70,700 today). He shouldered the staggering expense himself.[72]

The Touch and Delicate Thrill of Their Polished Surfaces

Randel's seven instruments were distinctive. He explained their design in a pamphlet that appears to be lost. Without that pamphlet, deciphering his diagrams is challenging, and by necessity incomplete and speculative, because his creations are so unlike contemporary instruments. One has to imagine holding them, carrying them, or pushing, pulling, and turning the elements. One has to live with the drawings, as instrument expert Jeffrey Lock graciously did for three days in the summer of 2010. He sat in front of an expansive computer screen, zooming in on details of scanned images of Randel's diagrams, analyzing, comparing the designs to other instruments he had studied. Because Lock restores and fixes Colonial instruments, constructing missing or broken parts with unsurpassed artistry, he knows surveying equipment down to the number and orientation of the threads wending round a tiny screw.[73]

At first glance Randel's instruments appeared to have a Rube Goldberg quality. After his first day with Randel's diagrams, Lock considered the disturbing possibility that the surveyor might have been "egocentric to the point of stubbornly going against common sense." Was it possible, Lock had asked, "that he carried that to an excess, to where it became counterproductive? To

SHORT MEASURING ROD

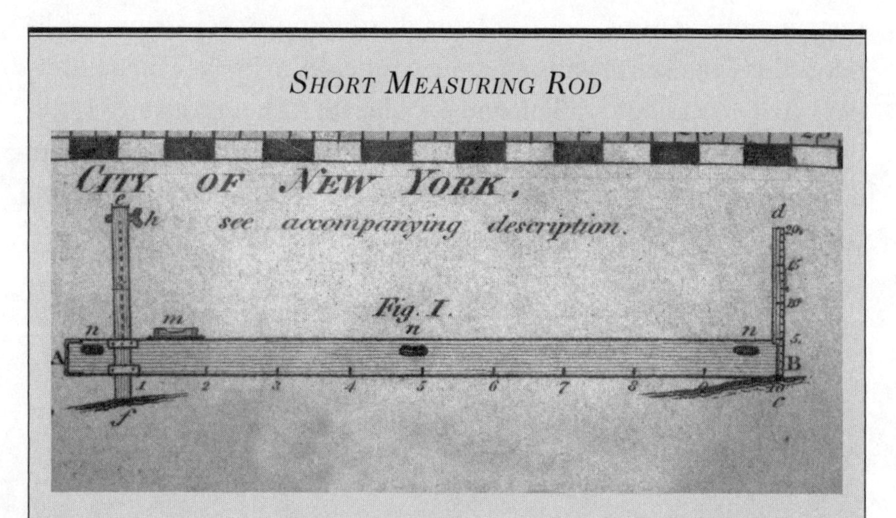

This long ruler was well suited to bumpy ground. The 10-foot metal shaft (A–B) was fixed to the ground at one end (B) and held securely in place by a vertical bar (d–c). (Most early nineteenth-century diagrams portray wood with wavy lines, metal with straight parallel lines.) On the other end (A), the rod could be moved up or down a vertical bar (e–f), so it remained level as the terrain changed. A level vial (m) indicated the even horizontal plane. Three handholds (n) made the instrument easier to carry. Randel had several of these 10-foot rods made.

where logic did not dictate his decisions?" Perhaps Randel had constructed crazy complicated devices when he could have just used straightforward existing instruments.

In time and with patient study, though, Lock began to see genius in Randel's devices, and he came up with thoughtful interpretations—a series of possible explanations for how the unusual instruments may have worked. In minute details, in the thumbscrews, the almost camouflaged springs, the micrometers and level vials, Randel's nuanced and mechanistic mind comes vividly alive. Careful sustained examination of Randel's diagrams yields interesting insights. For one, the instruments draw on and reflect his family's talents. They combine woodwork, metalwork, and the fine-tuned, delicate capabilities of a jeweler. For another, they reveal that Randel was making a painstaking survey of elevation as well as horizontal distances. In addition to achieving geodetic-grade accuracy, he obviously wanted to solve problems posed by hills, extensive acreage, a tight deadline, and an unsophisticated crew. His designs addressed all those concerns. "He has certain criteria for his instruments. One is portability. Two is ruggedness. Three is accuracy. And four is accuracy with flexibility," Lock sums up. "He had no problem totally redesigning something from the standard if it speeded up the survey process. Basically, he was a mechanical genius. He understood a lot of physics and he understood a lot of geometry. He had a lot of math training. He thought as an engineer. These are very logical, and functional."

IN PASSING FROM ARABLE TO GRASS AND WOODLANDS

Randel was proud of his instruments, touting their "peculiar construction" and the "scientific manner" they imparted to his surveying. He even claimed that "of course, no error has, ever

MEASURING ROD WITH COUNTER

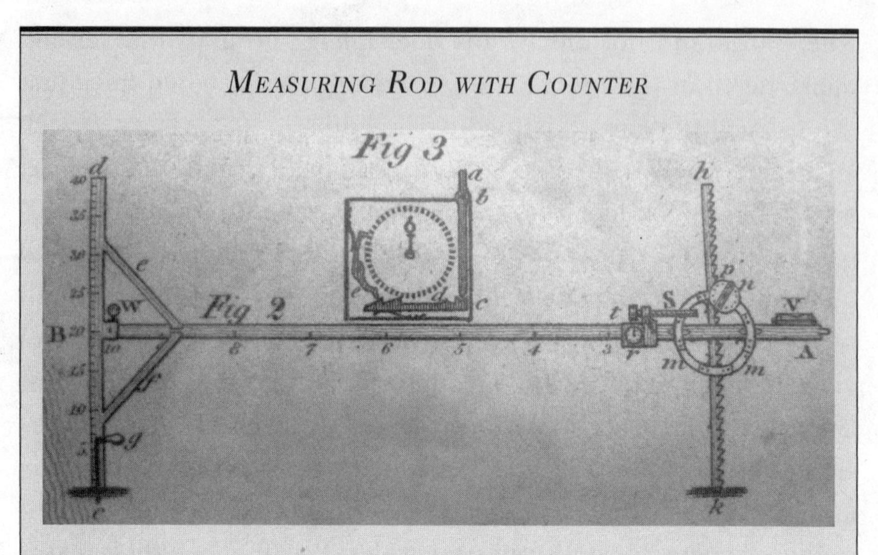

This 10-foot rod was created for ground where elevation changed more dramatically—for "exceedingly rough terrain," Lock says. The fixed end of the rod (B) kept a perfect level, because struts (e, f) held the horizontal rod in place; a level vial (v) was placed at one end. The other end of the rod (A) could be ratcheted up and down a toothed vertical bar (h–k). (It helps to think of the rod A–B as unmoving and the bar h–k as moving up and down with the terrain.) Randel designed a metal circle (S) as the ratchet; springs kept tension against the vertical and horizontal components (B–A and h–k), so the ratchet mechanism would not slip. In cases where perfect level was achieved only when the horizontal bar rested between two teeth, Randel devised a micrometer (r), "to fine tune between the teeth," Lock explains. "Think of it like a microscope—you have coarse and fine adjustment." Randel's use of larger teeth is unusual; most instrument makers of the time made only fine-toothed devices. Accordingly, teeth were often broken and damaged. Randel employed both. "He has come up with a method where he has got the strength and the flexibility and the accuracy all bolted into one design," Lock says.

Instrument 3 is probably what Lock refers to as an escapement counter, or a counting device. Just as guards stationed at a museum entrance click a handheld counter to record each visitor, Randel may have pressed the button of the escapement counter each time a rod was placed. At the end of the day, the counter could provide the total number of rods. By keeping track of rod tallies, the device freed Randel to focus on other field matters.

been found in any part of the whole work." He took great care of them, using cushions and trunks to protect them when they were moved. He chose to include them on his 1821 map; no other map from that era depicted surveying instruments in this fashion. And he publicized the fact that they let him measure twice as much each day, with two-thirds less error: "It is known, from experience, that a line measured twice with these instruments on such a field, will not in any case differ more than one inch in five miles." The instruments, or ones very like them, came to figure prominently in a bid he made to survey Albany. "The cost of inventing and constructing these instruments has been incurred, no person can use them except J. Randel & brother, he has not yet had any application to use them this season, if his native place reasonably accept the above offer they will derive the principal advantage from this invention. With them correct work is performed with an astonishing rapidity which enables him to work so cheap."[74]

Randel deserved to be proud of his creations. Most surveyors did not innovate by creating instruments that responded to the demands of a particular terrain or assignment; they used only what was available to them. Randel's resourcefulness and mechanical ingenuity remind Deborah J. Warner of the National Museum of American History of Scottish engineer and surveyor William Roy. Roy was leader of the British Ordnance Survey and is credited with originating modern geodetic standards and techniques. Warner says Roy's work, which was written and published in 1785 in the *Transactions of the Royal Society of London,* may have had a major influence on Randel.

Randel experimented with the effects of temperature and moisture on materials, just as Roy had. He tested mahogany, maple, pine, cypress, and metals. By immersing his rods in boiling water and then cooling them, he calculated that they would change in length 3 feet, 9.5 inches per mile between the extremes of summer and winter. He kept careful records of the temperature of the rods

LONG MEASURING RODS

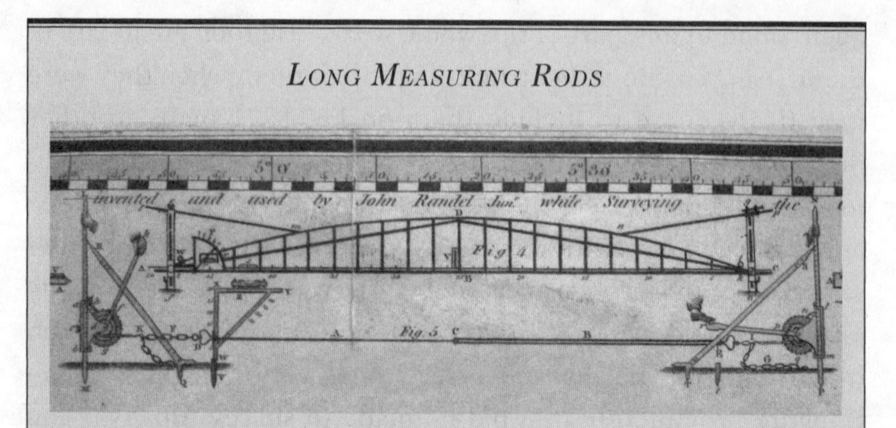

To move quickly over level ground, Randel devised additional instruments. Figure 4, a 50-foot rod, looked like a suspension bridge—and reminds Lock of an instrument described by Mason and Dixon to determine a degree of latitude. Randel's crew placed such rods end to end. Then, Randel wrote, the rods "were bolted to the ground; after which the end of the metal rods were brought in contact with each other, by means of fine tangent screws . . . the perfect contact of the ends of the rods was determined by the touch and delicate thrill of their polished surfaces." Much of his design aimed to keep the bar in place, and the level could be checked at several points (w, s, t). Quadrant (P) was an inclinometer crafted in the inverse of the standard design. Usually an inclinometer needle has a heavy point and hangs down from a rod, pointing to zero on a scale if the rod is parallel to the ground. As the ground rises the bar rises and the needle drops to one side like a plumb line, crossing the scale at the angle of incline. By contrast, in Randel's instrument, the entire bar moves, and the quadrant moves in concert with it. Somehow the needle remains fixed. "He might have found that if you hung the quadrant from the bottom of the truss rod it got damaged—the workmen hit it upon the ground," Lock says. "So what he has done is elevate it so it is safer. It works just as it would have traditionally, but it is not going to be damaged by being banged around."[75]

Randel created a collapsible rod too (Figure 5). Midway along one rod was a point (C) at which the rod to one side (A) might slide in between two others (B). Alternatively, C was a joint and the rod could be folded in half. Randel created a gear system to tighten and maintain tension on the rods and to ratchet them up or down to ensure the level. The triangle (w-x-y) supporting a level (z) poses an unsolved mystery.

when he used them: "The temperature of the rods was ascertained by placing and keeping the naked bulb of a thermometer upon each of them—the thermometer and metal rods being inclosed in thick pads of wool, to prevent the effects of sudden changes of temperature, in passing from arable to grass and woodlands." By knowing the property of his materials, Randel compensated for changes in measurement driven by temperature. He could calculate the difference between a distance recorded in the field in midwinter with a certain rod and a distance recorded in midsummer with that same rod. He could add or subtract seasonal variations in his materials so that his maps were based on a consistent measure of 50 degrees Fahrenheit, his average of the annual temperature. Doing this meant many extra calculations and extreme vigilance about temperature. And Randel didn't stop there. He also noted atmospheric interference and corrected for it by repeating measurements. "I was delayed by some undulations," he wrote one day when heat waves, water vapor, or refraction of the light interfered with his sights.[76]

Roy's account of his measurement of Hounslow Heath, conducted in 1784, seems like the kind of document that would have consumed and delighted Randel. One hundred and thirty pages of particulars about chains, metals, woods, glass rods, repeated experiments, and measurements of expansion down to 498 one-thousandths of an inch: "Hence it is evidence, that the dew imbibed only in one night, or space of time not exceeding fourteen hours, occasioned such an expansion in the deal rods, as in the whole base would have amounted to 45.484 inches," Roy reported. ". . . Here, indeed, lies the great objection to the use of deal rods." Music only to the ears of a similarly exacting, similarly obsessive mind.[77]

Roy himself was building upon the work of many scientists, including Jean Picard, Giovanni Domenico Cassini and his son Jacques, Pierre-Louis de Maupertuis, Charles-Marie de La Condamine, and Pierre Bouguer. These men had surveyed widely and

THEODOLITE

Randel's theodolite was unusual because it had two telescopes and because it was quite large by American standards. By estimating the usual width of a tripod leg at 2 inches in diameter, Lock reasons that the upper telescope was about 45 inches long. Both telescopes could probably pivot on the horizontal, so they could be set at right angles to each other if a right angle was needed. The top one could also pivot on the vertical to a greater extent than the lower one could; Randel could use it to record elevations (using the circular vertical scale, M–N). Alone of all the instruments, this one offers a clue to the maker. The scale (E–F) and the vernier (the smaller scale running from 15 to 0 to 15) for the upper telescope are read on an angle, or bevel. American instrument makers typically laid their scales and verniers flat. Only one of the two instrument makers Randel approached in 1819 imported English instruments and designed his scales on the English pattern: Richard Patten.

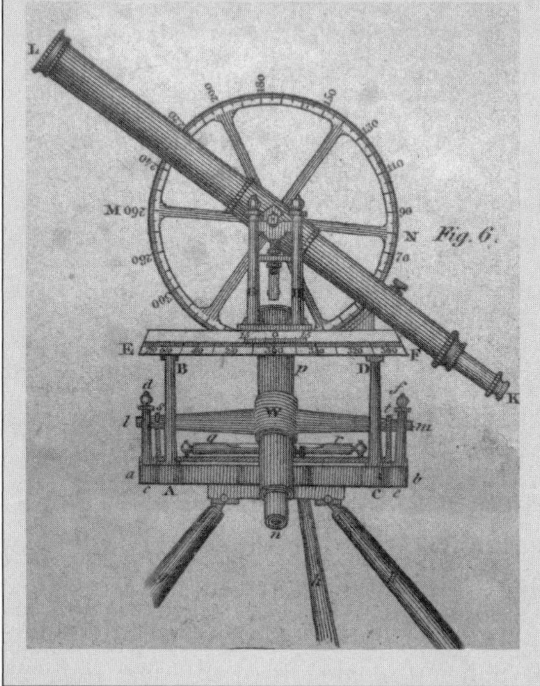

had debated one another across the seventeenth and eighteenth centuries, trying to refute or confirm Isaac Newton's hypothesis, put forward in the 1670s, that the earth was not a sphere but an ellipsoid—an oblate one at that, flattened at the poles, wide around the middle, more pumpkin than ball. The Cassinis had postulated the opposite—a prolate spheroid. They thought the earth was egg-shaped, flattened at the equator, pointier at the poles. The debate had become vitriolic, undiplomatic, a matter of national pride, pitting a Cartesian worldview against a Newtonian one. Living in exile in England, the French author Voltaire shifted his allegiance to Newton and ruthlessly poked fun at some of his countrymen. When Maupertuis, a Frenchman not on the side of the Cassinis, announced proof for Newton's claim in 1737 after a surveying expedition to Lapland, Voltaire joyfully called him "my dear flattener of worlds and the Cassinis." Even a similarly successful South American expedition, beset by disease and disagreement, foundering in mountains and rainforests, was not spared his sharp wit: "You have found by prolonged toil, what Newton had found without even leaving his home." The British and the French came together in the late 1780s to resolve differences in their respective locations of the Paris Observatory and the Greenwich Observatory. Survey teams led south by Roy and north by Jean-Dominique Cassini, grandson of Jacques, met in Dover in 1787, the year Randel was born.[78]

Determining the shape of the earth was one of the great geographical quests of the seventeenth and eighteenth centuries. Knowing the shape and size of the earth was, and remains, crucial to understanding distance and degrees of latitude and longitude, to giving coordinates real meaning. The key to those calculations was triangulation, which had been devised in the 1500s by Gemma Frisius and developed in the 1600s by Willebrord Snel van Royen, better known as Snellius. As the shape and curvature of the earth came into better focus in the 1700s, surveyors used spherical trigonometry to calculate how angles and sides alter when a triangle deforms across the earth's surface.

Scientists had discovered a great deal about the earth's dimensions by Randel's era—certainly enough to make highly accurate calculations of distance. Randel was familiar with spherical trigonometry, and considerations of the earth's curvature arise in his notebook. He also would have known about the controversies and the quest for the shape of the earth. He would have known that Newton had been proved right. But he would not have known the precise shape of the oblate ellipsoid whose surface he so loved to measure. That mystery extended well beyond the nineteenth century, requiring the development of instruments Randel would have reveled in. As Lock sees it, Randel and the contemporaries who shared his sensibility needed just a bit more knowledge, a few more discoveries, and then they "would have been the ones sending men to the moon."[79]

ALTHOUGH NO ILLUSTRATIONS attest to them, Randel may have devised instruments in addition to the seven described on his 1821 map. Several official documents mentioned a brass sector with a 54-inch radius that Randel used to take his elevations, both on a job outside New York City and on New York island. In one of those sources, the sector was used in tandem with a 30-inch telescope, which would have been unusually large for the time, Lock notes. And in one field book, Randel described a device to measure distance through rotations: "I have an iron wire 4/25 inch diameter of which I intend making a perambulator of 12 feet circumference, with which to measure the distance of fences & etc from the monuments."[80]

None of Randel's instruments apparently survived. A relative of Randel's reported in the early 1900s that he had donated Randel's theodolite to the New-York Historical Society, but no theodolite is listed in the collection now. It would have been an appropriate donation. With the instruments, Randel marked the grid, setting 1,549 monuments and 98 iron bolts that established the straight lines and right angles of the streets New Yorkers now inhabit. In 1817 Randel submitted a map showing the elevations of all the monuments along First, Third, Fifth, Eighth, and Tenth Avenues south of 155th Street. The map, now in the collection of the New-York Historical Society, shows elevations in cross section along these avenues traveling down the island.[81]

WATER LEVEL

Instrument 7 most intrigues Lock: "This is like it is from the planet Venus." It was most likely what was called a water level. Such instruments came to be used in surveying canals, in situations where extreme precision about elevation and grade was crucial to determining water flow. Randel seemed to have attached the telescope to the back of the ellipselike structure so it did not move. He would then, Lock suggests, get his general horizontal level by adjusting at the base of the instrument, using two screws (a and b). He would lock one of the vertical bars (G–F) into position with two screws at the very top (t and v). Once the coarse scale was set, Randel could make fine adjustments to get the telescope perfectly horizontal by adjusting the screws (f and d) and thus the movement of the other vertical bar (H). Using the micrometer (e), he could align the telescope by minutely shifting the entire ellipselike scale. "You are much, much more accurate this way because the telescope is locked into the entire triangular system, and then you adjust the telescope and the system and the scale all based on these thumbscrews," Lock says. "It provides rigidity." If Randel was recording water flow in the streams and creeks of the island, no record survives—and there seems to have been no request from the city government to gather such information. Randel may have designed the water level to establish a highly precise horizontal level, from which he could make obsessively exact angle measurements for elevation or decline.

The instruments also helped Randel achieve terrific precision with the block lengths, a detail about which he received no instruction. In the remarks accompanying the 1811 plan, the commissioners said they had specified the street width to be 60 feet and the avenue width to be 100 feet. The 1807 act, however, had been specific only about a minimum of 60 feet for the streets, and it does not seem to be known when or why the commissioners settled on 100 feet for the avenues, or how they decided on a few other details. For instance, the east-west block lengths between some of the avenues varied on the map; the distance between First and Second Avenues was 650 feet, yet that between Fifth and Sixth Avenues was 920 feet. And several streets, such as 14th, 23rd, 42nd, 72nd, 86th, 125th, and so on, were as wide as avenues. At some point the commissioners must also have specified twelve avenues. The grid was an act of imagination most particularly along First Avenue and Twelfth Avenue. Randel often found himself running those avenues through water: "36 street is in the River," "34 and 33 streets fall in the East River," "here stop, tide being in, cannot measure southerly over water."[82]

The particulars of the north-south block lengths seem to have rested with Randel rather than with the commissioners. The city left it to him to calculate them, which he did by averaging the distances between sets of streets to get a standard block length: that is, on average, about 200 feet. A block plus a 60-foot cross street totals 260 feet—the metric many New Yorkers recognize, and the one that makes twenty blocks roughly equal to a mile. (A mile is 5,280 feet, a near-product of 260, the same as 80 lengths of a Gunter's chain.) Uniformity weighed heavily on Randel's mind even in an early decision about blocks lying just *below* the southern border of the great grid at North Street—blocks not officially in the new grid system. In 1812 Randel proposed to the City Council that he divide a surfeit 11 385/1000 feet equally between the seventeen blocks that ran from First Street at the western side of North Street to Lewis Street on the eastern end. In other words,

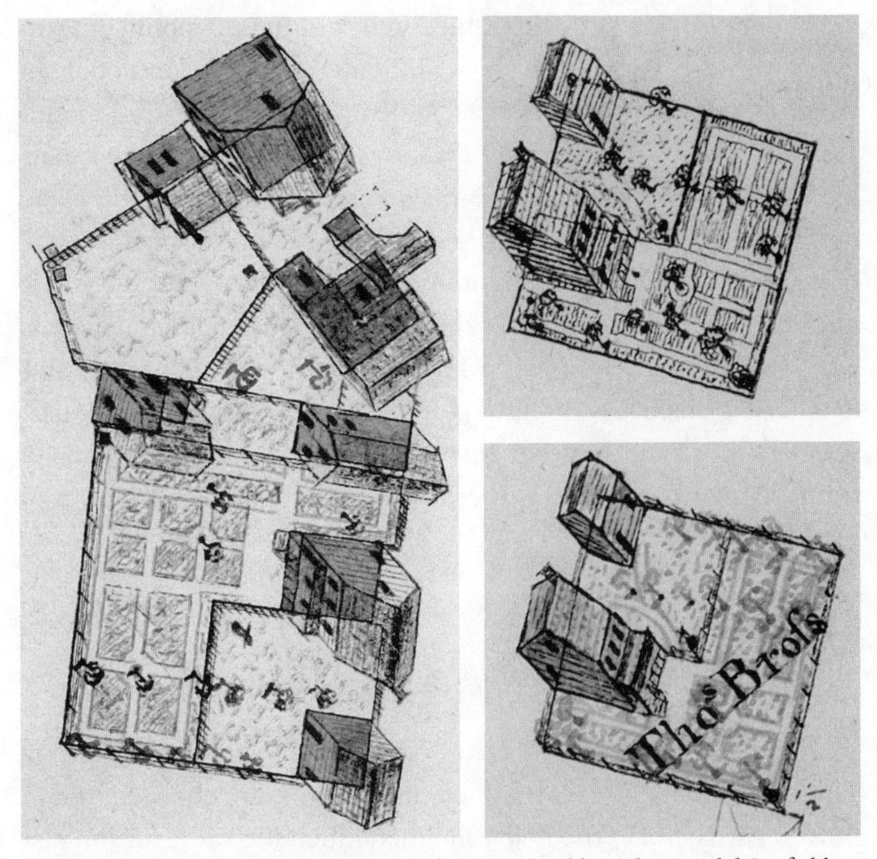

Drawings of houses and gardens on Manhattan Island by John Randel Jr., field book No. 66.3. Courtesy of the New-York Historical Society.

Randel wrote to the council to get permission to add 669/1000 of a foot (8.02 inches) to each block. Which he received.

He did not, however, consistently receive the Common Council's blessing for his obsessive exactitude. At one point Randel submitted a report to the council requesting $11,479.31 more for the extra work he undertook to correct for earlier errors—errors made before he developed his instruments—and for remeasuring he had conducted to ensure the utmost accuracy. The council was not amused by the request. Randel's job was to be accurate, so correcting errors didn't entitle him to further remuneration, the aldermen concluded.[83]

The aldermen did sympathize with a different point in Randel's petition: vandalism. It is difficult to find a notebook in which Randel does not record a stolen peg, a disappeared monument. Many irate landowners removed pegs or destroyed marble monuments. The markers disappeared almost as soon as he and his crew put them down. As he rechecked Seventh Avenue between 85th and 90th Streets in 1816, he noted not only that pegs had been destroyed but that replacements had vanished as well. Occasionally he knew who the vandal was: "The pegs at 1 avenue on this cross line are ploughed up by Frederick Reggis." Randel's work did have to be replicated several times, through no obsessive compulsion of his own. The council allowed him $4,000 more.[84]

Chapter of Accidents for Tuesday

The instruments, the techniques, and the mission changed, but in most regards, life in the field remained much the same for the rest of Randel's time on the island. His horse kept disappearing. "When we go for the horse, he is gone," Randel noted in March 1815. "William walks home & sends Nichols down Sunday who finds the horse in Leonard Street." Rather than staying in the city, he had rented a house in Harlem on the old Dutch church's property, at the intersection of the Kingsbridge Road and Harlem Lane, now roughly 121st Street and Third Avenue. The church was whitewashed, surrounded by elms and neat gardens, and when Randel was in Harlem, he attended Sunday services there. At times he and the crew camped out. Not surprisingly, Randel knew all there was to know about the gear—his "quantifying spirit" pervaded every aspect of his day. When dry, the tent weighed 20 pounds; wet, 35 pounds. The cot and tent together weighed 43 pounds. The buffalo robe (blanket), 8 pounds. All told, the gear was 71 pounds when dry, 86 pounds when wet. Watchmen took shifts watching the tent and the instruments.[85]

The 8-pound buffalo robe, among other things, was eventually stolen. On one occasion Randel dismissed a certain Benjamin Westervelt because he "desires to be discharged that he may return home to Poughkeepsie." The next morning Randel's chaise harness was missing. He offered $30 for the thief and harness, $20 for just the harness, and made inquiries at the livery stables, where he discovered that a man "from Poughkeepsie with a scratch on his face" had sold the chaise harness at auction. "This must doubtless be Westervelt," Randel concluded. He and William pursued Westervelt to Fishkill, a small town about 70 miles north of the city on the east side of the Hudson River. After Fishkill, William went on to Poughkeepsie, another 11 or so miles north. There Westervelt's brother claimed he had gone to Hartford. William engaged a constable in Poughkeepsie to make inquiries about Westervelt. There is no record in the field books of Westervelt's arrest.[86]

Not all thieves by any means, the men of the crew were consistently inconsistent. "Wheeler, Kerney and Cormer are beastly drunk; discharge Wheeler & Kerney and pay them at the rate of $10 per month; promise Cormer only $1 deduction if he continues." On another occasion, Randel paid a man extra wages not to drink: "Engage to give Frederick $25 per month for those months he drinks no spirituous liquor. $20 per month if he does drink any for six months from this day." Another was impudent to him and was discharged. Tempers flared: "Woodham this morning at 9 o'clock in a rage takes my field book and dashes it on the ground. Curses me and my instruments and work . . . says he will not do another day's work and goes off cursing."[87]

Others were lazy, of particular annoyance to the driven Randel, as is clear from his "chapter of accidents for Tuesday":

> Last evening when I went to Harlem with William he gave Sperry charge of setting monuments. Sperry attended to his

> business till night, then directed Mr. Mulligan to take instruments and etc. to particular places, or come home and provide wood. This he refused to do. If I should let this pass unnoticed many inconveniences would result to my work as it must necessarily happen that some of them must at times have charge over the others. If Mr. Sperry (who knows my regulations) refuses to mind Mr. Mulligan when I direct Mr. Mulligan to set monuments, I will discharge him and deduct his forfeitures. But Mr. Mulligan might not know that this was to be done, and also as it respects other things which he refuses sometimes to do at the house when requested. I now inform him that I should like him to be more obliging than formerly when requested to do anything. And not wait till I direct him to do every particular thing, but show a willingness to do what is to be done. I will not make him forfeit any thing for this last disobedience unless he repeats it again. Though my regulations say he must do anything directed by me or my director. To his credit, however, I must say he has done well today.

Randel couldn't help but add: "The other men do well every day." Randel ultimately fired Sperry for a series of offenses. Sperry took more than eight hours to bring supplies back from the city and nearly an hour to walk 2 miles to retrieve a signal flag. He burned holes in Randel's carpet. He was, Randel wrote, "very slow and impudent and withal profane."[88]

Wood and Clarkson, two recurrent characters, tried Randel's patience as well. Not only did they take refuge from inclement weather in taverns, they were notoriously unreliable. Randel recounts a tale about Wood and Clarkson's journey to the city to get meat, a nail rod, and a stand. Wood was on foot, Clarkson in a chair with William. "Three o'clock they returned without having attended to one single instruction." Randel detailed what the two men did, or rather did not do, in the city. William must have been his eyes and ears, because Randel was home calculating distances. Wood and Clarkson didn't get the nail rod. Someone else brought the meat home. Clarkson hired a wagon to carry the stand "though it is not an inconvenient shape and weighs almost too little to mention," and then with "much injury of the stand, which was directed to be carried by hand, he arrived with the wagon and stand 3 o'clock." Randel concluded: "This delay has much the appearance of a determination to get here too late to carry monuments and their refusing again to day has the appearance of stubbornness or laziness."[89]

Although tension stirred, Randel seemed to strive to treat his men fairly and well. When he celebrated his birthday on December 3, 1812, he also had to lay off men for the season, and tried to do so in the best possible way. As he recorded, "This morning I conclude that I cannot employ the nine men . . . all winter, as I need only three to work to advantage. I inform them all of my intention, offer to retain them till the last of next week so as that they may have sufficient time to look out for other employment and also (before I make my selection), to let those who desire to discontinue, or who are unwilling to engage now for another year (as I consider it to my profit to retain only those who will be with me next year) go, and to pay them their wages without any forfeiture as if they had remained till April. Give them till this evening to conclude."[90]

RANDEL'S FAMILY VISITED regularly. His father, mother, sisters, brothers, and various cousins and other relatives stayed with him or stopped by on their way back and forth between Albany and Orange and Bloomfield. Although his field books record constant work—walking, measuring, calculating—Randel stopped his grueling pace one weekday to accompany family and friends on a carriage tour of Manhattan Ville, Harlem, and Hell Gate. He accompanied various friends and family to various ferry landings. But he does not seem to have socialized much outside his family; whenever parties are mentioned in the notebooks, it was usually William who attended.

Randel possessed a keen sense of familial responsibility. He consistently employed William, who was a loyal aide, but who also left Randel's notebook out in the rain, made surveying errors, and may have had a problem with drink. William was fiscally irresponsible at times as well. Randel served as guarantor for some loans William was to repay with interest. When William did not repay the loans or provide interest, Randel had to sort out the debts, sometimes ending up in court for them, as happened with a promissory note William gave to Abraham Riker. William counted on Randel for work during nearly every surveying job Randel took, and when Randel did not have a contract, he paid William to bring wood, or $3 a week to make breakfast. In an 1823 letter he wrote that he could not aid William at the moment, that he didn't know what work he would be doing next, but that he hoped to get a contract on a New Jersey canal, "which if I can get will give you considerable work and ready money." He offered to let William have some upstate land for a third of its value, payable in ten years with 6 percent interest. "As nothing can be done however without cash and you desire to have a farm that you can call your own, I will aid you all I can," he wrote.[91]

Randel also routinely sent money to his brother Abraham, who in 1813 had started a farm in Sconondoa, now Verona, New York. The land was in Oneida County near Oneida Castle—a village Randel had laid out, which was at various times proposed as an alternate state capital to Albany. Randel "considered that a great development would be sure to come to this section, and he counseled his brother, my grandfather, to come here and buy land at once. The capital did not come," recalled Randel's great-niece. Also among the first Verona settlers was Samuel Breese, a cousin of Samuel Breese Morse, who invented the telegraph. The Randels were close with the Breese family, and Randel often sent his regards or asked that items be delivered via Breese. Abraham established the first Presbyterian church in Verona and built a three-story house in the Albany style. "In the rear of the house, on the banks of Oneida Creek was an Indian Trail, and many interesting stories have been told about the friendliness of this tribe," recalled Abraham's granddaughter. "Often when my Grandmother was working about the house, the door would open softly and she would hear a voice saying 'Sagola,' Indian for 'How do you do?' She never felt at all afraid of them, and they always borrowed her dishes to prepare their meals on the banks of Oneida Creek. And their request for 'knodelute' (bread) was never denied. We still have a very large iron kettle at home, which they frequently borrowed and which was their special joy. The trail was a much traveled one and often our family was lulled to sleep by the sounds of their Tom-toms." The house that Abraham built was recently razed, but the road remains Randel Road.[92]

Randel purchased several parcels of land upstate for himself and on behalf of his family and acquaintances. Two of the many lots that appeared in his records were Lot 74 and Lot 71 in Verona on a tract called the First Pagan Purchase, formerly belonging to the Oneida Indians. Both lots were surveyed in 1809 by James Geddes, a judge, politician, and civil engineer who moved to the region in 1793, helped set up the lucrative salt industry in Onondaga, and later directed work on one leg of the Erie Canal at Simeon DeWitt's request. Lot 74 had nearly 140 acres was a long rectangular property, one end angled and bounded by land belonging to the Oneida Christian Party. "The soil of this lot is good," Geddes wrote. "The front fine wheat land . . . The back part for corn or grass." And indeed, Randel and various assistants did plant corn and wheat, and dug a well and cleared stones. Lot 71 covered a little over 125 acres and was bounded by Oneida Creek; the road to Rome, New York, passed the property to the south.[93]

Whether Randel himself or Abraham settled on Lot 71, Lot 74, or another parcel is not clear. But the brothers did establish farms near each other. Randel ran his from afar by hiring caretakers when he had the money, and he appeared intensely involved in decisions about the land, the crops, the livestock, and the neighbors'

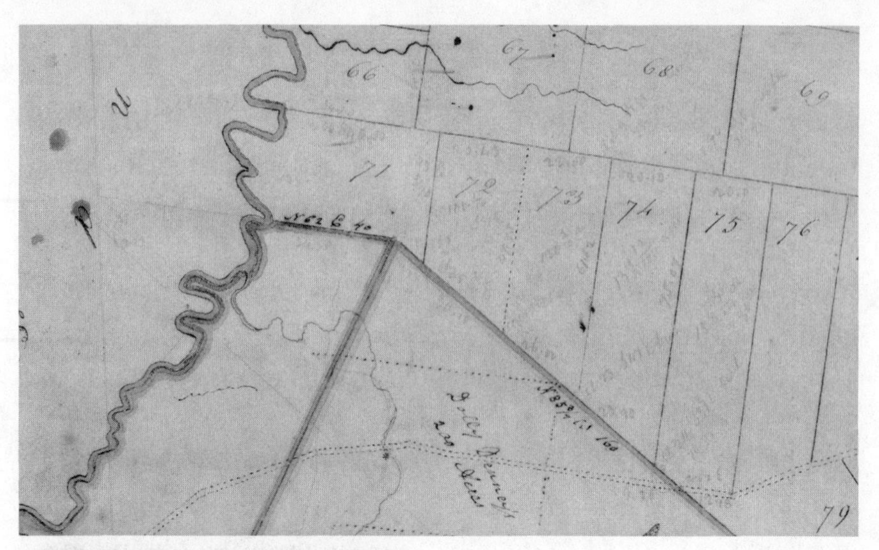

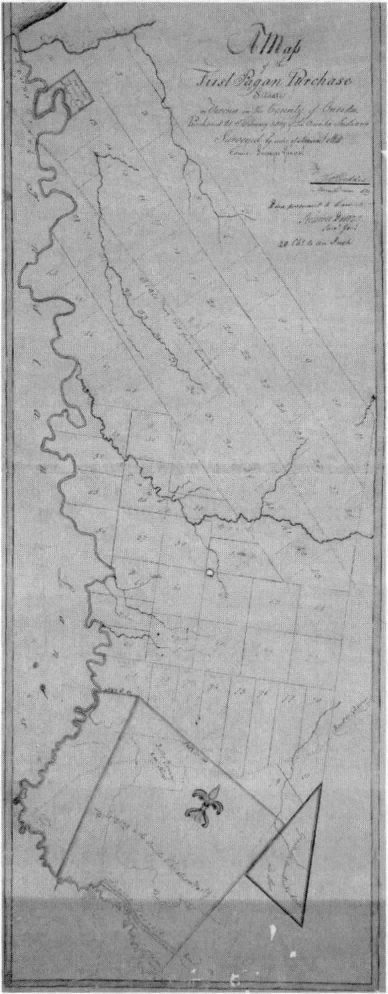

Randel owned lots 71 and 74 in the First Pagan Purchase, which became Verona and was surveyed by James Geddes. Map No. 44, Series A0273. Courtesy of the New York State Archives.

cattle when they trampled his trees. He tapped maples, using at least forty buckets, which Abraham also used or stored for him. He grafted apple trees; five hundred are noted in one letter. He carefully chose his pigs and made their travel arrangements. "I have not had any letter from you respecting the sow. I therefore have concluded that you could not find any boars of the right breed," he wrote in a draft letter to Moses Morris of Cranberry, New Jersey. "I have concluded to take in addition to the 2 sow $2 boar pigs, I spoke to you about which are now not far from six weeks old, one other boar pig of the same litter and four sow pigs of the no bone breed (which you had in a pen behind your house) which are now say about 4 months old. As Mr. Simpson will not be at New Brunswick, I agreed with Mr. Christian Van Nortwich Tavern Keeper in New Brunswick to receive the pigs from you and pay you for them—he will direct them for me, to the estate of Mr. Wm. Brown. No. 53 Broad Street New York who will receive them for me and pay the captain of the packet boat for Mr. Van Nortwich."[94]

Randel was particularly concerned about one horse, one he had discovered mired and bruised during a visit. "I am pleased to hear that the blind horse is doing well. Indeed I was almost certain that he would do well because he was under your care and you knew that I valued him highly," he wrote to Frederick Bryant, either a neighbor or a caretaker. A month later he wrote, "Do not stint the blind horse—although I made no exception to selling every article of produce and fodder I did not intend that he should be neglected." We do not know which brother was more successful as a farmer, but Abraham and Randel were frequently in contact about farming. Randel served as his brother's agent for selling hops in New York City. He helped Abraham buy hogs, locate seeds and tree cuttings.[95]

Although Randel was consistently employed, his fiscal burdens began to grow after 1810. He was not reimbursed for most of the cost of his instruments. He was not reimbursed for the extra time he put into remeasuring the city once he had his instruments. He had to advance his men's wages. He had to lay out for monuments, bolts, paper, ink, instrument repairs, travel, and any other expenses before he was reimbursed, which sometimes took a long time. At one point, unable to meet payments on a house he had bought in Albany, he appeared quite overwhelmed: "The purchase of the house has been a most unfortunate thing for me . . . for a person of my age and situation is almost too much." He wrote to his bankers, "When I purchased this property I hesitated for fear of sickness or accidents happening to prevent me complying with your conditions," and quoted their earlier agreement not to "crowd me nor take advantage." He said he had considerable money coming but couldn't collect it at the moment.[96]

When unable to meet debts on his Oneida farm, Randel instructed his caretaker not to give Abraham any more grain, because he needed to sell it all himself. But, he begged the agent, "As to brother Abraham however I wish you to use all the tenderness on my account." On a different occasion, he told Abraham that he himself owed money, that he couldn't get more until his work in the city was done, that he worked every night until twelve or one o'clock to get that work done, that he had tried to borrow but couldn't, and had then tried on mortgage but had failed in that too. "What money I can send I will and you after paying your debts must do the best you can."[97]

Randel expressed love and affection for his family, and he must have felt responsible not just because of his Presbyterian upbringing, which emphasized familial and social duty as well as good moral conduct and aid to the needy, but because he recognized his position as the most successful and well known of his siblings. Perhaps because of this success, he came to view himself and his family as worthy of the trappings of affluent families: the several farms, tracts of land, houses. Randel also commissioned portraits of some of his family from Ezra Ames, a famous Albany portrait painter. Ames painted the Albany elite, including Simeon DeWitt, Gouverneur Morris, George Clinton, and the wealthy businessman Elkanah Watson, and was a member of the same organizations Randel belonged to, the Mechanics Society and the Society for the Promotion of Useful Arts. It was Ames whom DeWitt commissioned to paint the eclipse of 1806. Whether driven by aspiration or by a sense of rightful belonging to high society, Randel invested heftily in the oil paintings. Records from Ames's account books show that Randel ordered a portrait of his father, his sister

Portrait of John Randel Jr. (left) and portrait of Abraham Randel, likely by Ezra Ames. Photographs courtesy of Sotheby's, Inc., © 1980.

Hannah (by her married name, Mrs. William R. Weeks), his mother and sister Jane DeWitt, Abraham, and himself. The portraits cost between $25 and $35 apiece, and the burnished gilt frames $21 or so each. Randel must have spent at least $250 on the paintings. Today that would be about $3,640.[98]

Portraits of Randel and Abraham turned up at a Sotheby auction in 1980, listed as done by an unknown hand. Nevertheless, it seems certain that Ames painted them. Not only do Ames's record books list Randel as a client, but the style is consistent. In those paintings Abraham looks much older than his brother, who was younger by only a year. He is facing the painter, his arms gently crossed, his hair gray. Randel looks at the painter too but is turned slightly to one side, holding a roll of paper, likely a map of New York City, although it is hard to make out in the catalogue photograph. His white collar is larger than Abraham's and looks more flamboyant, without actually being so. He is dark-haired, handsome. He looks strong, a bit stocky, intense, and somewhat mischievous.[99]

Mrs. Randel Joins Me in Respects

Randel's portrait captures him at just the time he was married. Whether Randel fell in love with his cousin Matilda, whether she fell in love with him, whether the family urged a union between cousins familiar and friendly with one another, may never be known. Although the Randels and Harrisons and their extensive families appear to have written frequent letters, none between Randel and Matilda has emerged. But Randel's field books and letters reveal Matilda as a true partner, one of usual "good spirits." She accompanied him on surveying trips upstate, even during difficult weather. She assisted in one survey near Oneida and may have helped on one in Manhattan. (If this interpretation of the notebooks is accurate, Matilda could have been one of the only female assistant surveyors of her time.) And when Randel worked late into the night to meet deadlines for New York City, Matilda helped him create beautiful maps.[100]

Randel and Matilda were married in a joint wedding ceremony with Abraham and Matilda's cousin Rebecca on May 11, 1813, at the First Presbyterian Church in Orange, New Jersey. (One twentieth-century family account describes Matilda and Rebecca as sisters, but several New Jersey and other genealogical records identify them as cousins.) The couples were married by the Reverend Asa Hillyer, with whom Randel and Matilda maintained a close relationship for many years. Randel bought land on Hillyer's behalf in upstate New York, facilitated his payment of taxes on that land, and informed him about family matters.

Rebecca, who grew up not far from Matilda, proved a good partner for Abraham. She was intrepid, traveling with him on horseback from Albany to settle in Sconondoa, in a part of the state "which to them at that time seemed a wilderness," where she planted a grove of maple trees. She and Abraham had seven children, and her granddaughter Ogdena Fort remembers stories about her. "One time Rebecca got on the slow boat at Durhamville to visit relatives in Utica and Albany," Fort recalled. "She took her old rocking chair along, fearing that her relatives' furniture would not be comfortable. When the boat reached Utica, the Boat Captain carried her rocking chair and a lantern, to where the relatives had lived previously, only to find they had moved. It was raining, but Grandma Randel was so tired she asked the Captain to wait for her to sit in the chair a little while, so after she rested, they proceeded to the new address. Mrs. Randel spent the night there, and in the morning took a fast sloop to Albany. The Captain's feelings about all this are not recorded."[101]

No such anecdotes about Matilda survive, but details about her family and information about women in New Jersey and the Presbyterian church in the early nineteenth century reveal something of her background. Matilda was born in 1789 to Aaron Harrison and Phebe Crane. The Harrisons had been among the first settlers of New Jersey and held prominent positions as farmers, politicians, and church members. Aaron Harrison had first been married to Jemima Condict. Condict, who lived from 1754 to 1779, kept a diary during the Revolutionary War that is well known to some scholars of women's history and the early American republic. "It seams we have troublesome times a Coming for there is great Disturbance a Broad in the earth & they say it is tea that caused it. So then if they will Quarel about such a trifling thing as that What must we expect But war & I think or at least fear it will be so," she wrote on October 1, 1774. She chronicled unrelenting death and sickness, her thoughts about marriage, her intense religious feeling, and self-criticism: "This Day I am eighteen year old the Lord has been So Mercyfull to me as to spare me so long When I have bin sining against him Dayly, Sins without Number. O may I Resolve to Spend the rest of my time in fear and in thy Service." Condict also showed independence and humor: "Thursday I had some Discourse with Mr. Chandler. he asked me why I Did not marry I told him I want in no hurry. Well Said he I wish

I was maried to you. I told him he would Soon wish himself on maried agin. Why So? Because says I you will find that I am a crose ill contrived Pese of Stuf I told him that I would advise all the men to remain as they was for the women was Bad & the men so much worse that It was a wonder if they agreed. So I scard the poor fellow & he is gone."[102]

Condict died before Matilda was born, but her diligent records of sermons and her harsh judgment of herself portray a family ethos, particularly the ethos of rigorously Presbyterian families. Matilda's father was a deacon in the Orange church, and Matilda must have attended church every Sunday. She would have learned the skills to run a large household and to take care of an extended family. She would have been frugal, but she was willing to be less so: "Matilda has three yards of cloth she got of her mother for a coat. It is coarser than she wishes. I take it at $3.50 and she will get finer cloth," wrote Randel in 1819.[103]

Presbyterian women did not have a formal role in the church, where they were expected to be subordinate. In the domestic realm, however, they were expected to be the guardians of morals and of children's education. "As long as they acted through persuasion rather than authority, women were promised extensive power, particularly over children, and even over patriarchs," writes Mary P. Ryan about Presbyterian women in *Cradle of the Middle Class*. In the early nineteenth century, Presbyterian women became more active in their churches and communities; some began forming societies and reaching out to the larger world. Notably, they began to support missionary and charitable work, collecting money and clothes and forwarding them to foreign missions and to the southern United States. One of Matilda's relatives, Mrs. Matthew Harrison, was directly involved in such an effort, and a series of letters in one of Randel's field books chronicles the disappearance of a trunk sent by the young ladies of the town of Preeble in Courtland County. The trunk had been entrusted to Mr. George Dyckman, who was to deliver it to John

Sayre, a member of the American Board of Commissioners for Foreign Missions, an early U.S. mission organization. The charitable work of women, and young women in particular—part of the religious revival movement that came to be known as the Second Great Awakening—stirred tension. Randel's letters noted that if the trunk were not found, it would give support to those "not favorably disposed towards Missionaries."[104]

Surveying trips and outreach to missions on behalf of Presbyterian women were likely as close to engagement with the sphere outside the home as someone of Matilda's era, age, and background would get. Matilda would not have had a chance to participate in the unique, short-lived right of New Jersey women to vote. During the Revolutionary War, women in that state had often found themselves in the thick of battle. "Numerous New Jersey women were forced to flee their homes or see them and themselves plundered. Following the battle of Springfield, the *New-Jersey Gazette* reported that 'six widows are burnt out; some very aged, and others with small families; and almost all the houses in the neighborhood which were not burnt, were torn to pieces, entirely plundered.' Hanna Caldwell of the renowned Newark Ogdens, a mother of nine who chose to stay in her home, was shot," according to historians Judith A. Klinghoffer and Lois Elkis.[105]

For various reasons—recognition of women's patriotic participation during the war, arguments for women's rights—the New Jersey legislature interpreted liberty liberally in the new republic. All residents of a certain age and worth (50 pounds in the first post-Revolutionary version of the state constitution), black and white, male and female, could vote. In practice this meant that wealthy widows could vote, as married women could not own property. In 1807, when Matilda was eighteen, women and blacks lost the vote; the Federalists and the Republicans each agreed to give up a constituency that had supported them. The reasons were complex and had to do not only with biases and state politics but with haggling over a site for a new county courthouse.

Apparently Newark (near Bloomfield and Orange, the seat of the Harrisons) initially objected to the reversal. Whether the Harrisons were among the dissenters is not clear.

Matilda joined Randel in Harlem after their wedding in 1813 and worked closely with him for the duration of their marriage. A few weeks before their wedding, Randel had his crew bring to Harlem his guns, equipment, a chest, furniture, and a portrait. He instructed William and two other men to set up the garden. Matilda brought to the Harlem house a sideboard, table, stand, and work stand—perhaps part of her dower, or perhaps the house needed more furnishings in order to become a home. She kept the home and garden, traveled with Randel, ferried back and forth to New Jersey to see her family, entertained and cared for visiting relatives and friends. Randel wrote of her frequently in the notebooks, noting how she was and where she was, what she had done. He always sent her regards or respects or love along with his own when he wrote letters to people they knew jointly. Their first year together seems to have been uneventful. Randel continued to measure and calculate so he could place monuments and bolts in exactly the right place. The calm was broken in early 1814.

Taken from the Map Made by Me

When Randel began his New York assignment, he was young and untested. Within two years he distinguished himself as a man of accuracy and was consequently awarded a significant high-profile contract and a chance to design his own instruments. A sense of self-righteousness, fueled by his fixation on mathematical exactitude, seems to have taken hold during this time. Perhaps it had always been there. In 1814 Randel wrote to a local newspaper, seeking to redress a wrong, protect his reputation, and drum up

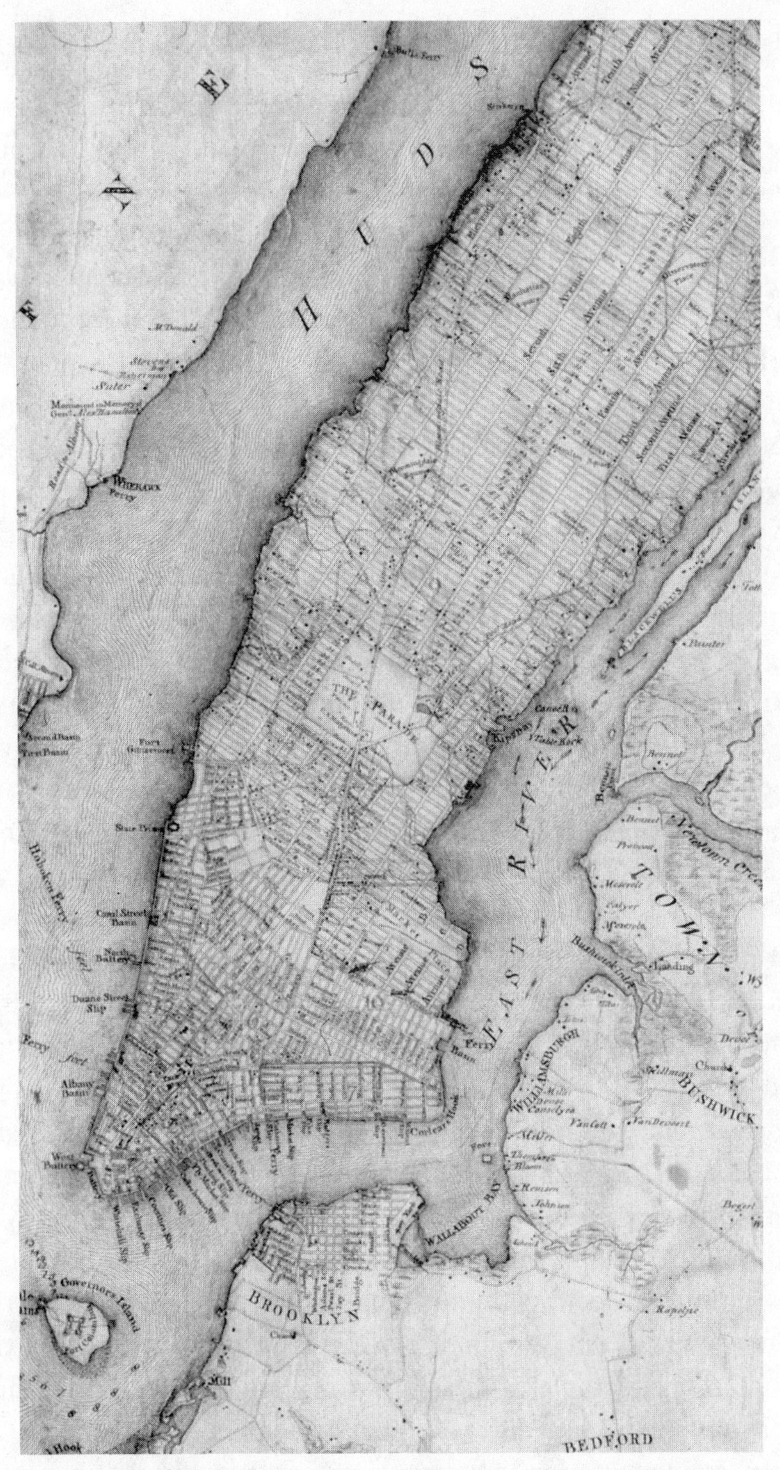

Detail of the 1814 Commissioners' Plan by John Randel Jr. Courtesy of the New-York Historical Society.

public support for his position and a commercial venture. It was the first time he had done such a thing. It was a pattern he would repeat for the rest of his life, with wildly uneven success.

The conflict had its roots in 1811, when Randel delivered his copies of the Commissioners' Plan. Just a few weeks afterward, the city surveyor, William Bridges, a man of mixed reputation, petitioned the Common Council for the copyright of the plan. He said he intended to engrave the map and give a free copy to each member of the council; he also intended to obtain subscribers (in other words, sell it for his own profit). The Common Council gave Bridges permission, noting that anyone had the right to print and publish the public map. Bridges borrowed a copy of Randel's map in May, and in November his version, rendered by master engraver Peter Maverick, appeared. Bridges had 342 initial subscribers, including a John Randell—although Randel later claimed he had never purchased one. Randel's name is nowhere on the Bridges map, which is today the best-known version of the Commissioners' Plan and a copy of which is housed in the New York Public Library.[106]

If Randel chafed at Bridges's cartographic appropriation, he did not express it until three years later, when he prepared to sell his own reproductions of an engraved version of his 1811 map. Randel's 1814 maps were to be in color, 22 by 34 inches, and to cost $2.50 for those who preregistered to buy them, $4.00 for those who did not. (The respective prices are about $33 and $52 in 2011 values.) Randel felt a need to differentiate his map from Bridges's earlier one in order to increase sales (to cover, as he put it, the expense of "plate and paper") and to establish his authority, accuracy, and authenticity. He published in the *New-York Evening Post* on March 21 an advertisement about the pending sale, complete with supporting documents. Without initially naming Bridges, he stated that his forthcoming map "will be found on examination to be more correct than any that has hitherto appeared, and that part of it which contains the plan of the city cannot be made more accurate . . . As the author was Secretary and one of the Surveyors of the Commissioners appointed by the state for laying out the Island, and in that capacity possessed all their materials, and as he has since compleated the measurements and fixed monuments by contract with the honorable Corporation, he alone is possessed of all the materials for this valuable work."[107]

Just below his advertisement, Randel reprinted a copy of a letter from himself to Gouverneur Morris, to whom he had submitted his new map for "inspection" and "candid opinion." Randel then shared his own candid opinion with Morris: "I have compared Mr. Bridges Plan, (selling at 8 dollars, in sheets, not coloured) with that which the Commissioners reported, and which, as their Secretary, and Surveyor I made.—It is far from correct, indeed (excepting a few alterations lately made in the city) the only things in it which are accurate are taken from the map made by me, and yet he has ventured to claim the work as author, and as such has obtained a copy right for what is in truth the copy of a public record." Randel included Morris's response just below:

> I have examined the map sent with your letter of the 10th as fully as my other engagements would permit. I have seen it with great pleasure and consider it as an excellent work. It is, in a manner indispensable to those who wish to make themselves intimately acquainted with the Topography of that interesting space which it comprizes. It appears to me more accurate than anything of the kind which has yet appeared. Indeed until all your actual measurements were compleated, it was hardly possible to attain to that accuracy which the totality of the materials in your possession has enabled you to exhibit. Without entering into questions which already exist or may hereafter arise between you and Mr. Bridges I shall have no hesitation in recommending your work. I consider it as highly deserving of public patronage.

Randel also reproduced a pithy statement from the street commissioner attesting to the map's accuracy.[108]

Whether Randel intended to trigger a public feud or not, he did. Bridges quickly and vituperatively responded in the *New-York Evening Post*: "Observing in Mr. Randell's advertisement of his Map of Manhattan Island a direct and illiberal attack on *my* Map, for the purposes of extolling his own, and enabling him to meet the expences of 'plate and paper' I feel compelled reluctantly to say a few words on what I consider unprincipled, and most assuredly unprovoked conduct." Bridges referred to Randel's "cloven foot," to his deficiency of both "candour and honorable conduct." He asserted that the commissioners had signed his map and that Morris did not say that "*my* Map is wrong, but being Mr. R's patron he says he will recommend *his* work. That very answer conveys a hint which would instruct any person less conceited than this young Man, that it is as essential to his reputation to be correct in his manners as in his Maps." He ended his letter sarcastically, citing the "astonishing accuracy to be alone found in Mr. Randell's practice."[109]

Randel countered with a long, passionate, detailed response, suggesting he had been aggrieved for some time. His letter ran in the *New-York Evening Post* a week later. Apologizing that neither he nor the man he addressed were of "sufficient importance to claim the public attention," Randel nevertheless demanded public attention because of the "value of reputation." Perhaps because of his upbringing and the shame of losing one's good name in the Albany Presbyterian community, and perhaps because he knew his future professional life depended on his reputation, Randel wanted the record set straight. He also wanted his first major work credited. He printed a letter that he said had been delivered to Bridges a week or so earlier, to which Bridges had not replied. Randel asked the public to peruse this letter so he might defend himself from obloquy, a then-popular word for abusive language or ill repute. He cited professional rivalry: Bridges had coveted the contract Randel had won from the city to lay the monuments. And he called Bridges out on the matter of the signatures: "You know that the Commissioners *did not* 'give their signatures in attestation of the accuracy of the manuscript of that part of *your* map,' but their signatures and seals are attached to the manuscript of *my* Map now in the Clerk's office, and for reasons which I shall presently shew, they never would have given their signatures to your Map."[110]

Then, with his emergently famous precision, he enumerated the flaws Bridges had made when he copied the map. Randel had quietly been collecting evidence of Bridges's professional inadequacy for several years. In one field book covering 1810 and 1811, he repeatedly made note of discordant findings: "Note W Bridges mark is 6½ inches E of my line" and "W Bridges mark is 2¾ inches E of my mark." Now he let loose with his analysis of Bridges's map. Islands were too short, too narrow, in the wrong place. Rocks too far north and of the wrong size. Hills misplaced, misshaded. Rivers too close together. Fifty-eight buildings missing, forty-five in the wrong place south of 155th Street, two forts too close together, and forty-two buildings in the wrong place north of 155th Street. "In addition to all these errors *you have* 'without a special act of the Legislature' closed part of the 26th and 27th streets, and also part of the 1st avenue." He included a lovely line, "You charge me with youth as if it was necessarily attended with ignorance," and went on to describe his long training under DeWitt. He said he could provide Bridges with a full list of mistakes instead of the few examples he had given, but that he had "no wish to display the above facts to the public." He would prefer that Bridges "spare me this necessity by retracting so much of your publication above referred to as is injurious to me."[111]

Bridges had stolen before, as map scholars and dealers Robert T. Augustyn and Paul E. Cohen note. There was nothing unusual in this. Many, if not most, mapmakers stole; copying from earlier maps was and remains an oft-necessary aspect of cartography. Randel did it when he needed to include areas he had not surveyed himself, but he always cited whom he had borrowed from and relied upon. Map scholars refer to the practice as "cartographic synthesis." It can provide important clues to cartographic lineage and can explain the persistence of mistakes, such as some of the state boundaries in Abel Buell's map, famous for being the first map of the United States drawn and printed in the new country. Erroneous maps were common in Colonial America and the early republic; in many cases, no replicating or rectifying surveys had been undertaken, and mistaken depictions of geographical features and boundaries persisted until surveys set the record straight. As a city surveyor, Bridges knew New York well, but rather than drafting his own maps based on his knowledge and experience, he relied on others' maps when he published, thereby adding and perpetuating errors. In 1807 he published a copy of an 1803 map made by Joseph F. Mangin, one of the designers of New York's City Hall, and Casimir Goerck. (Goerck died in 1798, shortly after he did his survey for the Common Council, and so the map was mostly Mangin's work.) The Common Council had hired Mangin and Goerck to make a map of existing homes, property owners, streets, important buildings, and so forth—a state-of-the-city map. The surveyors, who began their work in 1797, proved to be less than scrupulous. They straightened the shoreline, molded irregular blocks into quadrilaterals, and created streets, including Mangin Street and Gorrek Street. Bridges seemed unconcerned with the inaccurate portrayal. When asked by Dr. Samuel Mitchell to contribute a map to his guidebook, *Picture of New York*, Bridges redrew the Mangin-Goerck map, noting in the title that it was "drawn from an actual survey by William Bridges, City Surveyor." No mention of Mangin or Goerck.[112]

Bridges had also been accused of impropriety before, over a matter noncartographic. In early 1808, letters in the *New-York Commercial Advertiser* accused him of interfering with an election, first by threatening a carpenter with unemployment if he didn't vote for the Federalists and then by trying to buy a vote.

Bridges's rebuttal appeared on May 7, 1808: "I understand that this *gentleman*, Peter M. Dustan, says that he heard me promise the dollar to the said negro near the Poll, when I was on one side of the street and he on the other. I leave the public to judge of the probability of my offering a bribe near the Poll, in a voice so audible as to be heard across the street." In March, the Common Council fired Bridges, which he insinuated was retribution for his Federalist politics in a city strongly Democratic-Republican: "The subscriber having been removed by the present Democratic Corporation, as City Surveyor, informs the public that he nevertheless continues to act as a sworn Surveyor, and shall execute any commands with which he may be favoured, with diligence and dispatch." Bridges was rehired as a city surveyor in April the following year. Payments to him then appear regularly again in the minutes of the Common Council, with references to his surveys, including one of a doctor's botanic garden.[113]

Although used to defending himself and to hustling for income, Bridges seems to have let the feud fade after Randel's letter of April 1814. He may not have had the heart or energy to fight. His wife, Phillis, was ill, and died in June. A few weeks later Bridges himself died, aged forty, leaving seven children and many creditors.

Despite the passionate and heated exchange, Randel did not publish or engrave his 1814 map. Only a manuscript version was produced. As Augustyn and Cohen note in *Manhattan in Maps*, Randel was concerned about national security and about issuing a map more up-to-date than the 1811 Commissioners' Plan. The misleadingly named War of 1812 was still under way; it officially ended with the Treaty of Ghent in December 1814, but it did not end on the ground until the Battle of New Orleans, in January 1815, during which the British troops, who had not yet received word that the war was over, sought to claim the lands of the Louisiana Purchase. Between 1806, when the British had begun attacking American ships, seizing men and goods, and

1815, battles had been fought on many fronts, including in and around New York City, Washington, D.C. (where British soldiers burned the Capitol), the Great Lakes, and the Chesapeake Bay. America had done much to secure its future in the early 1800s, but few Americans could have felt secure about their new republic as the British parried and thrust up and down the coast, and Randel was no exception. "The Public are respectfully informed that the Map of New York Island and its vicinity, prepared by John Randel, Jun was in the hands of the Engraver, and would have been out by December next, conformably to his engagement, but it has been suggested to him that under present circumstances it might be improper to furnish the enemy with an opportunity to procure by means of his agents such accurate information of the country—he has therefore taken it back and will postpone the publication to a more proper season," read a notice in the *New-York Evening Post* on October 5, 1814.[114]

Postponed, but not forgotten. A constant perfectionist, Randel kept working on the map for several more years, expanding, revising and correcting, updating. He finally released an engraved map in 1821, at the very end of his stay in New York City. But it garnered him no greater satisfaction than the 1814 map had.

Impetuous Streams

Although still deeply engaged with his work on Manhattan, Randel became increasingly active upstate as the decade wore on. He frequently traveled by steamship, stagecoach, and horseback to Albany and central New York. Steamships greatly reduced travel time, making surveyors and engineers more able to do several jobs at a time. "Formerly the passage from New York to Albany was considered an affair of a week or ten days—three days was called good, and forty-eight hours excellent—though a fortnight

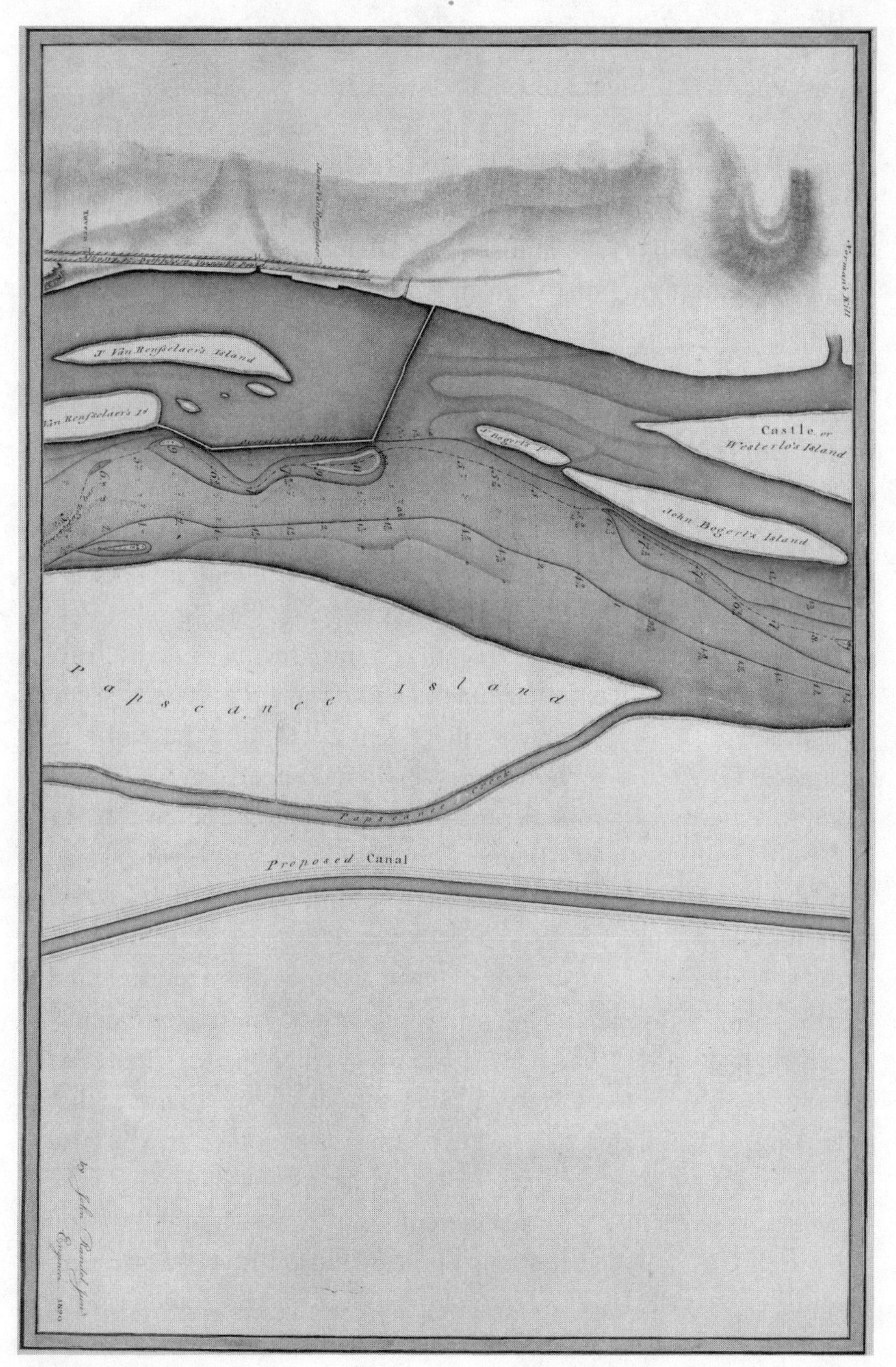

Detail of Hudson River map by John Randel Jr. New York State Surveyor General; Maps Produced for the Commissioners to Report a Plan for Improving the Navigation of the Hudson River, 1820–1830. Series A4006-78, No. 7. Courtesy of the New York State Archives. (Albany and the north are to the right.)

was not very uncommon," a contemporary passenger wrote. "Now, however, the same voyage is currently made in thirteen hours, sometimes in twelve . . . What would good old Hendrick Hudson, the original founder of the colony, have said, had he looked out of his grave and see our gallant steamer, the Constellation, come flying past him like a comet at the rate of twelve knots an hour?" However, once delivered by steamship, passengers had no easy voyage across land. "I could not hire a wagon to go for any money the Roads were so exceedingly bad," Randel wrote to DeWitt once when stranded in Salina. He opted to go by "saddle horse" and "kept behind some large wagons that cut the trees that the wind had blown down."[115]

Between 1811 and 1822, Randel surveyed and mapped in Oswego, the New Military Tract, the Onondaga Salt Springs Reservation, and Oneida. On one trip, nine Oneida Indians traveled with him to mark their houses and crops on his map. "They discover no anxiety about a village being laid out," he wrote to Simeon DeWitt, "they however will claim compensation for their improvements." He took depths in and mapped the Hudson River for a stretch south of Albany. Commissioners and businessmen paid for the survey. They were interested in constructing a canal from Albany south 16 or so miles to New Baltimore, so as to avoid the ship-deterring shoals and shallows. Randel had experimented with a way of determining depth when he was working in Oneida. "For that purpose I had a large square of wood made, suspended a plumb bob and went to work," he wrote to DeWitt. "I found that looking with the eye over a board the refraction or rough edge prevented my observing the object with any tolerable degree of accuracy, particularly as this difficulty was increased by the wind blowing." He finished the maps of the Hudson River in the winter of 1819, when the river was frozen over and he couldn't leave Albany. "Our river has been closed upwards of a week. I crossed it twice today on foot, it is not strong enough for horses yet," he wrote to Asa Hillyer.[116]

Randel surveyed islands in the St. Lawrence River, where he became quite ill. "I have improved my health very little if any since I left the St. Lawrence," he informed DeWitt. "I have been taking emetics to remove the inflammation the Doctor thinks is on my lungs and cathartics to carry off the boil that was collected in considerable quantities. Last week I was so much better as to run one of the Village Streets, this week I scarcely go out of the house, though from the quantity of phlegm thrown off by the emetics, that I am in a fair way to get well soon."[117]

For all the sparseness of Randel's records, injury, illness, and disease are richly documented. He sprained his ankle. He cut his thumb badly, "but it has not interrupted my work any, except a few hours when I had the wound sewed up." He had boils so extensive he could not ride. William had boils so extensive he could not work. Randel had ague. William had ague. Crew member Smith had ague and fever. Randel had fever. He had "a pulmonary consumption." Randel walked so fast he got "so much heated as to have been in danger." He had headache for four days, with pains darting through his temple and the back of his head. "Have tooth drawn." "Ran a nail in my right foot and am lame." Matilda had a swollen face. Matilda "continues sick with a cold." Matilda had dropsy. "All of us are confined home with colds." An associate and his children have "been sick with the bilious fever." "Mr. Gilchrist has a thorn in his hand." A friend's father died of apoplexy. Sister Rebecca's husband, Robert Dodd, died of consumption. The "usual autumnal fevers" arose. Randel stayed away from Syracuse for fear of getting a pervasive illness: "I have learned that it became very sickly immediately after I left Syracuse. Dr. Kirkpatrick I have heard has been dangerously ill. I shall return there as soon as the health of the place will make it prudent so to do."[118]

The litany of ills reminds us how vulnerable people were to every aspect of disease or infection back then. Exactly what these various ailments were remain unclear. Historical references to smallpox and measles during this era are often accurate, because the diseases were so distinctive, says James Colgrove, a professor in the Center for the History and Ethics of Public Health at Columbia University. But other illnesses were fungible. "The naming of disease in the first half of the nineteenth century was loose and inconsistent," Colgrove notes. "They didn't differentiate. General fevers would be called different things, and different doctors would call them different things." Ague is a case in point: it may have been a particular disease or just a common fever.

One illness Randel experienced in the winter of his thirty-first year is particularly noteworthy. He suffered a liver ailment, or that is what his doctor or doctors diagnosed. To treat it, Randel cycled through the standard, and as he described them "severe," treatments: bleeding to clear out the bad blood, blistering so the site of the burn and resulting sore would release the illness, and hefty doses of mercury as a laxative. "The early nineteenth century was the age of heroic medicine, of doctors like Benjamin Rush," Colgrove explains. "Doctors were very, very aggressive in giving mercury and mercury in high doses. It is horrifying to read some of the accounts of treatment. If [Randel] had been treated extensively with mercury it is not at all far-fetched to think it affected his personality." Mercury, a heavy metal, was known to have adverse effects in Randel's day. "Mercury, when administered by a skilful hand, is a most valuable and salutary medicine," stated an advertisement that ran in many newspapers at that time. "On the other hand, a rash, indiscriminate and unqualified use thereof among mankind, has been productive of infinite mischief—our present limits forbid us attempting an enumeration of the evils that have arisen from the abuse of Mercury." By 1860 such evils were recognized to include mad hatter's disease. Mercury is toxic and deadly at high doses. In adults it can cause motor problems, timidity, misanthropy, and diminished intellectual capability, among other effects, none of which seem to have affected Randel. Mercury can also trigger rapid mood swings and quarrelsomeness, two behaviors that do seem to increasingly characterize Randel from his thirties on.[119]

Randel's reports of illness and injury were more frequent upstate, where fieldwork took him into remote terrain. His challenges in the field in central New York were often greater than they were in Manhattan, as one account to DeWitt illustrates:

> It is now between 1 and 2 o'clock in the morning, and I have been up since an hour before day yesterday so as to get my chain men on the ground rather before light. The Roads are very bad. The rains having taken away the bridges in several places, required us, sometimes to go round a considerable distance, and at others, when there was no way round, to ford the creeks and the impetuous streams nearly overset us. The last 3 hours it has snowed and we now have 3 inches snow, on the soft mud. I intend if practicable to get to Onondaga by the night of the 14th (or this tomorrow night) and then have these papers mailed so that you may have them by the 16th inst. I intend when I get to Onondaga, to get assistance as you recommended if needful.[120]

Randel's work upstate differed from his Manhattan work in another significant way. Away from the city, Randel was not always inscribing someone else's blueprint; he had more freedom. In Oswego, for instance, DeWitt instructed him to use his discretion in determining the size of lots. This gave him the chance to display a topographical sensitivity he was unable to express in New York City. In Oneida Castle, he reported to DeWitt, he tried to get the streets parallel and at right angles to either the turnpike or another road but had not succeeded. "The plan in either case would have appeared handsomer on paper, but would have cut the ground much less convenient for building lots than

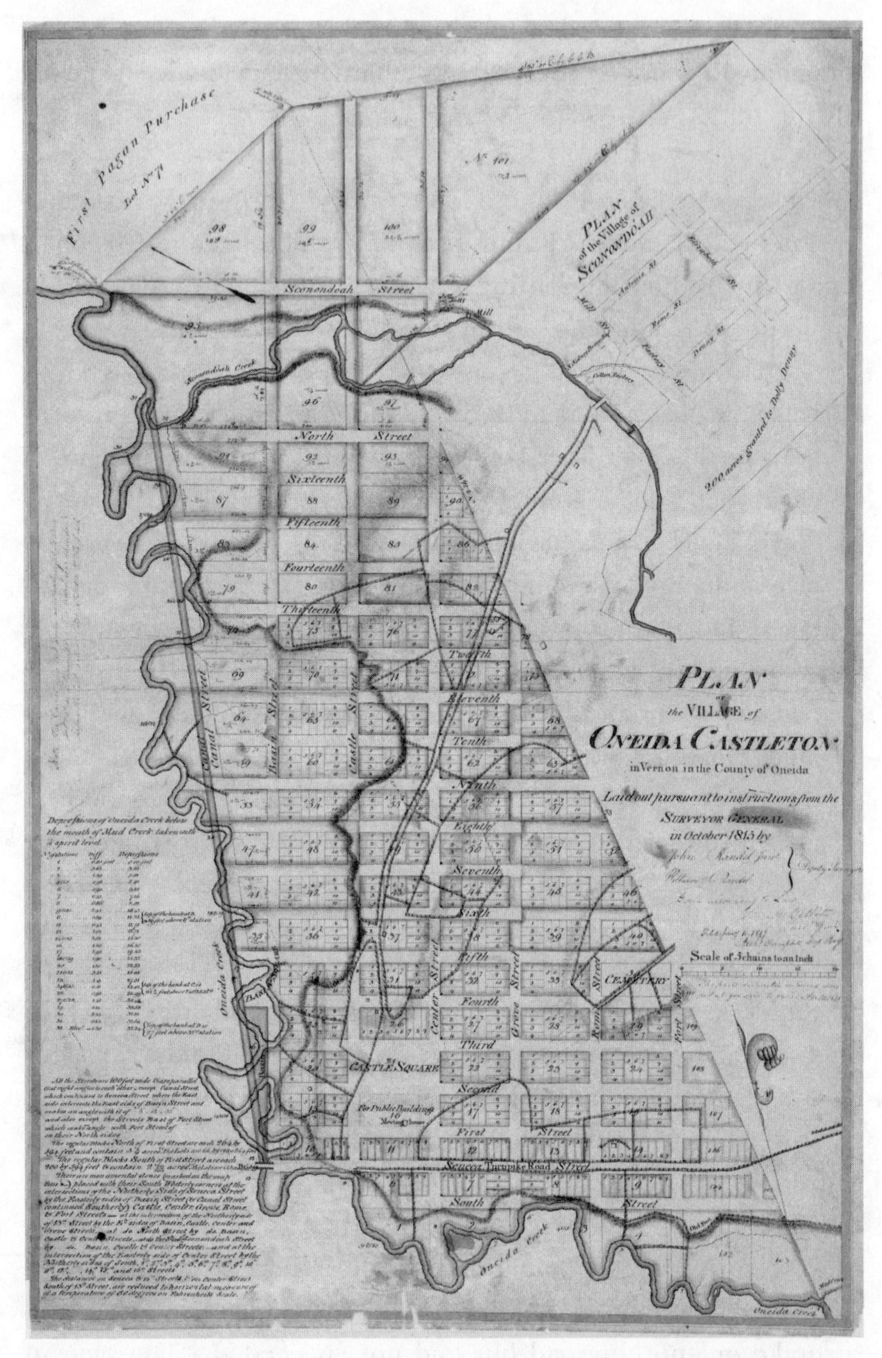

Map of Oneida Castleton by John Randel Jr. Courtesy of the Bureau of Land Management, New York State Office of General Services in Albany.

the present. To remove one objection to this Plan I have made a triangular public Square or place, where it is most proper for the Streets to bend." With usual affection, he signed his letter, "Mrs. Randel joins me in respects to Mrs. DeWitt and Sister, my love to the boys and James."[121]

Randel had another vision for Oneida Castle too:

> In laying out the Public Places which you have directed to be at the Meeting house, I thought as this spot was the highest it would be best for Public Buildings, and the gently descending ground south of it would be best for a Public Place. I have reserved 2 blocks for Public Buildings and as the ground southerly of the South reservation is of little value, it will make a pleasant front or rear perhaps for those buildings. I have selected a level, dry, loose soil for a Cemetery, and so distant from the turnpike as to be in no danger of needing to be removed hereafter. With such a good road to it that persons may be buried with ease any weather. Although it is healthy at Oneida now, hereafter it may not be so, it is pleasant for families to have each a small spot to themselves. Like the famous Burying place at New Haven.[122]

These examples of Randel's initiative are significant. They indicate a strong, mutually respectful relationship with DeWitt and reveal an awareness of urban planning and design beyond New York. The cemetery he mentioned at New Haven is most likely the Grove Street Cemetery, founded in the late eighteenth century, which combines a grid plan (Enlightenment rationality) with an emergent parklike quality (nascent Romantic movement). In Oneida Castle, Randel sought to situate buildings and squares according to elevation and topography, which suggests that he may not have adopted a uniform grid in all parts of Manhattan if he had contributed as planner to the commissioners' vision.

It is a tantalizing possibility. Randel clearly valued the grid plan; a year before he died, he wrote a passionate defense of it for the longtime Common Council clerk David T. Valentine, saying, "This Plan of the Commissioners, thus objected to *before its completion, is now the pride and boast of the city*." Randel went on to note its "beautiful uniformity" and its other positive attributes: fire prevention, air circulation, enhanced real estate values. But his upstate work suggests he could also see a city conforming to the land's contours and shape it thus—as does his backing of a voluble opponent of the grid.[123]

Clement Clarke Moore, the owner of an estate called Chelsea (from which today's neighborhood gets its name) and declared author of "The Night Before Christmas," vehemently opposed the grid plan. In 1818 Moore published a pamphlet, *A Plain Statement Addressed to the Proprietors of Real Estate in the City and County of New York*, in which he complained about many aspects of the plan, including its topographical insensitivity. "Our public authorities seem unwilling to depart from the leveling propensities, but proceed to cut up and tear down the face of the earth without least remorse, and, apparently with no higher notions of beauty and elegance than straight lines and flat surfaces placed at angles with the horizon, just sufficient to suffer the mud and water to creep quietly down their declivities," he wrote. And it was Moore who penned the oft-cited line that the commissioners were men who "would have cut down the seven hills of Rome." Moore was in large part angered by the city's decision to flatten hills in his neighborhood, which would alter natural watercourses and disrupt sewage flow. (The Commissioners' Plan was never specific about leveling. It presented a map of the city on a flat plane; anchoring that flat plane or not was the city government's business.)[124]

One would not expect Randel's name to appear as a witness for the plaintiff in this document, but it does. Moore wrote that he and some Greenwich Village residents had consulted Randel about the leveling plan and that he supported their opposition, showing that the grading would lead to the flooding of several houses and gardens. Moore wrote,

> A tract of land large enough for a town to be built upon it, and already containing many houses, is doomed by the corporation of the city to be raised several feet, in order to make it coincide with a plan devised by their street-commissioner. The owners of this tract, seeing themselves likely to be grievously oppressed by this plan, employ a person, of confessedly the first rate abilities and of much experience, to examine it. He pronounces the plan to contain regulations absolutely unnecessary, and ruinous, not only to the tract in question, but to the land beyond it; and declares that, by such a mode of regulating the ground, no calculation can be made of the ultimate mischief that may ensue. He proposes several plans in lieu of the one objected to, and urges his objections with a mathematical precision to which the corporation, with all *their* counsel, have not been able to oppose a direct answer.

The plans Moore described were not alternatives to the grid; rather, Randel offered several alternative plans for Moore's neighborhood, including leaving the hills as they were. City government, as Moore notes, disregarded Randel's alternatives.[125]

Randel was not an uncritical booster of leveling and lines. Much as he loved straight rules and right angles, he was sensitive to topography, to curves—a critical sensitivity for his later work as a builder of canals and railways.

Matilda Pastes No. 87 for Me

The same year Moore issued his pamphlet, Randel began a project that rivals the Commissioners' Plan in its importance yet has remained oddly obscure. He undertook to make an enormous map of Manhattan on a large scale, a map that appears to have no counterpart. This masterwork of cartography is not just breathtakingly beautiful; it contains invaluable information about New York City's social and natural history. Randel's marshes look wet and filled with rushes. The orchards are lined with trees, the shorelines rocky, uneven, deeply blue. Familiar names echo: Dyckman, Astor, Beekman, Clinton. The huge map comprising ninety-two smaller ones, captures the city and the landscape in the early nineteenth century in a way no other does. This remarkable map, this little-known masterpiece, almost didn't come to be. It was an accident of persistence.

In November 1818, Randel reminded the Common Council of an overlooked agreement from 1812, from just after he had begun monument setting. That year he had proposed to the council that he make a detailed map, or maps, of the island at a scale sufficiently fine to show "Hills Valleys Rocks Houses Creeks" and to show "possession fences and lines of Real Estates." His proposal had been accepted—commissioned, in Randel's eyes—but the aldermen had apparently forgotten about it six years later. Upon receipt of Randel's reminder, the committee in charge peevishly rebuked its predecessors for not communicating with the council as a whole before entering into contracts. Nevertheless the Common Council honored the contract, and Randel set about creating ninety-two maps at a scale of 100 feet to an inch. Placed together, the 32-by-20-inch maps make up the largest known map of the city, stretching 50 or so feet. According to map experts Paul Cohen and John Hessler, nothing similar exists in the United States for that era. Hessler, senior cartographic librarian at the Library of Congress, notes that one equally detailed map was being made at the same time—in Japan. Surveyor Inō Tadataka (1745–1818) recorded every last detail of his island, at his own expense, taking some seventeen years to do so. The scale is not as fine as Randel's, but the vision and impulse seem to have been similar. Tadataka's maps, finished and published after his death, portray all of Japan and the intricacies of its coastline.[126]

It took Randel two years to complete these ninety-two gorgeous maps. He started in early January 1819, when he and Matilda retired to Orange and Bloomfield, New Jersey, so he could work on the project, taking with them "cakes and fruits for the children and beer." They arrived to find all in "tolerable health." Those months have stretches of near-daily notations in the field books, which capture more fully than any others the rhythm of life and work, and of travel around the region by horseback, wagon, or pleasure wagon. Excerpts from January show the close-knit family, the routines, the integration of work and home life:

- January 3. "Sabbath. Matilda and I partake of the sacrament in Mr. Hillyer's church. No addition to the church but one woman this time. Snows in the afternoon and evening."
- January 5. "Remain at Orange mapping till 12 o'clock. 13 hours."
- January 6. "Stop at Bloomfield. Find all in usual health, except sister's child, who is sick. Father has been making trials of map varnish . . . Cross and go to Mr. Rutherford's, where I dine . . . Return to Bloomfield and take tea at Orange. Stop at uncle Abijah Harrison's and set an hour. And get home to Father's 8½ o'clock. Commence mapping and continue three hours."
- January 10. "Hear Mr. Hillyer preach from 96 psalm and 2 Corinthians."
- January 11."Map till 1 o'clock 15 hours."
- January 12. "Finish rough copies of 9 maps No. 84, 85, 86, 87, 88, 89, 90, 91, 92, and commence fair copy of No. 92. Work till 12½ o'clock 14 hours."
- January 13. "Continue copying No. 92. Shade spitting devil and Haerlem Rivers to do which takes 8 hours. Father comes over from Bloomfield. Mother and sisters are in usual health. Sister's child is improving in health. This evening father

returns to Bloomfield in a wagon of Father Harrisons which is going with Fanny and Liddy to see their children in Bloomfield. Brother Charles wife and child take tea here this afternoon. This morning the back room was cleaned and white washed and new bedstead put up. Weather mild and fair."

- January 15. "This day Matilda my wife is 30 years old . . . Matilda goes with me to Newark."
- January 20. "Go to paper mill and get 25 sheets paper before this I got 4 sheets in all 29 sheets. Stop at Maverick's find him at my plate. Bring Sisters Catherine and Sally Eliza to Orange on a visit. Map 3 hours."
- January 21. "Misty weather. Take tea with sister and M at uncle Abijah's."
- January 25. "Brother Charles buys 13 candles for me. This evening observe that a depth candle will burn 6 hours and a mould candle . . . will burn 8 hours."
- January 23. "Sister in law Jemima has taken a cold and is in bed. I map all day No. 89. 7 hours. Prick off No. 88 and work on it till 11 o'clock. 5 hours. Matilda pastes No. 87 for me. Sun shined and it was pleasant and warm as April. Sat without fire all afternoon. This evening sky is overcast."
- January 29. "Map 8½ hours. Take tea with Mr. Hillyer at Brother Charles Harrisons yesterday and today clear and cold. Uncle Amos sent the map to Mr. Simpson by Cousin Bethuel."
- February 1. "Map 1 hour and finish the nine maps and start with them for New York."[127]

Randel seemed to be happy in New Jersey, despite the heavy workload. Matilda's uncle Abijah Harrison and he seemed to have got on well, given the frequency with which Abijah turns up in the notebooks. Abijah read widely, and he loved clocks. He had smuggled one past the British during the Revolutionary War "in the dark of the moon" to avoid the high taxes collected on clocks then. And his home, where Randel and Matilda often took tea, was cozy and well known in the neighborhood. The entrance hall had a mahogany sofa, with feet carved into alligator form, and a rocking chair sat next to the fireplace in the sitting room, where Randel must have "set."[128]

The family relied on Randel to sort out delicate issues. One of Matilda's elderly relatives approached him for help, claiming she had been defrauded by her sons. Randel's sense of responsibility and right propelled him to defend the widow. "I am really sorry for the old lady," he wrote to Rev. Asa Hillyer, updating him on the case, for which Randel had to engage a lawyer and get sworn testimony. "The mere mention of her receiving one cent less than her dower should have filled the mind of an upright executor with horror." Randel's involvement brought him into conflict with Matilda's brother, Aaron Burr Harrison, who at times surveyed with him. "If I had not obtained the affidavit, I should have been under the necessity of ruining my brother Burr's testimony by the testimony of his two sisters, Jemima and Abbey. Under this impression, I concluded I could save him by an exit." All seems to have been resolved, and Randel's intervention was evidently supported by Matilda, at least two of her sisters, and her father, Aaron.[129]

Randel often worked past midnight to finish the maps. As he chronicled in his field book, he would first make a rough copy, sketching out the map (based on sketches from his field books, his measurements of distances, and his notes on elevation, various structures, and property owners). He would then lay the rough version of the map over another, clean sheet of paper. Using a sharp pin, he would prick outlines from the rough sketch through the two pieces of paper so tiny holes traced the correct features on the fair copy. He would ink features and then paint or shade the fair copy, which would then be pasted to a muslin back and varnished. Unlike the sketching and tracing, the varnishing went quickly: Randel coated thirty-five maps one day. Matilda did at

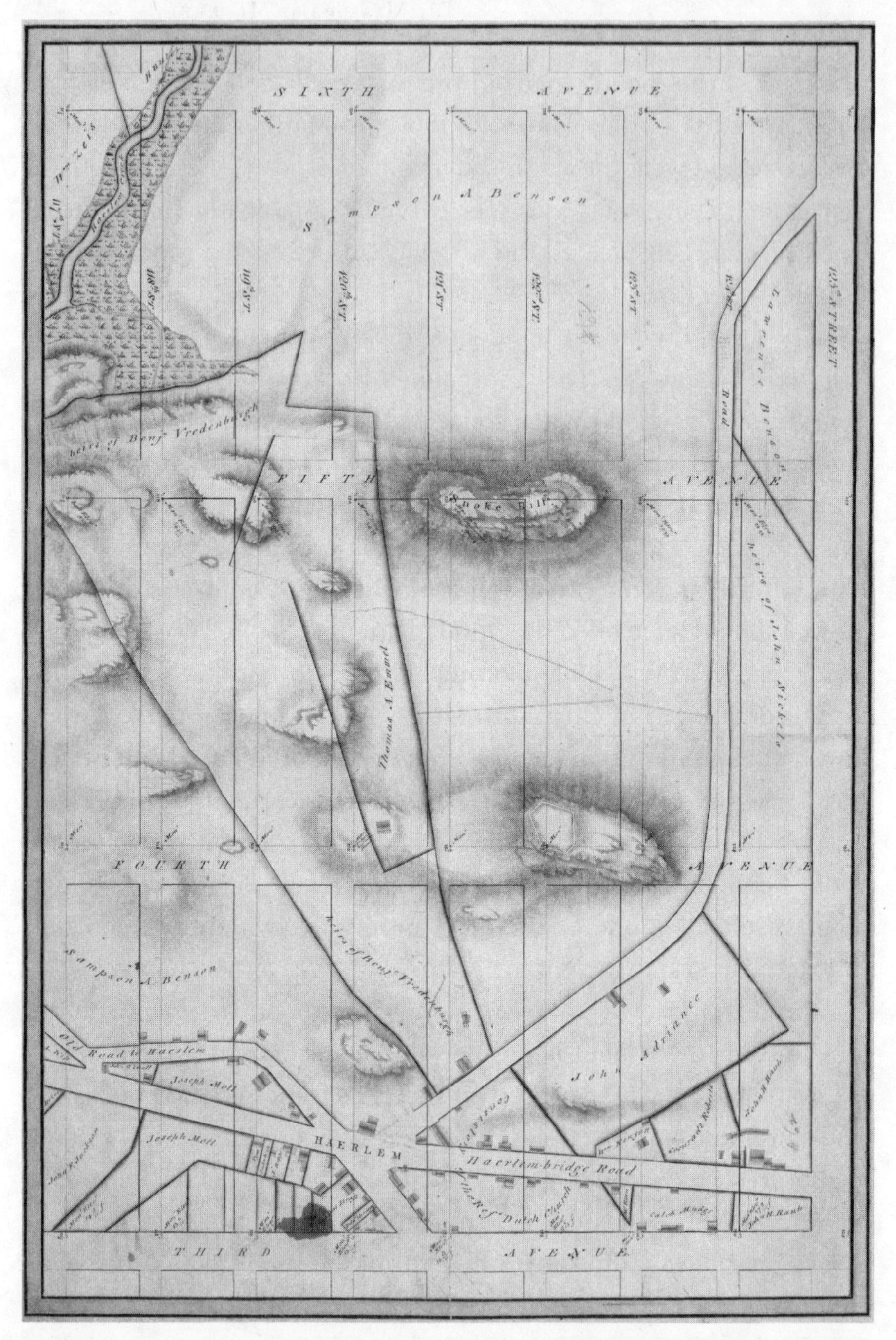

Farm Map No. 62: Harlem, the land of the Reformed Dutch Church where Randel lived, and Snake Hill (Marcus Garvey Park today). Randel drew, and then bound together in four volumes, the maps with avenues running left to right and streets running top to bottom—an orientation peculiar to New Yorkers today. Courtesy of the Manhattan borough president's office.

least 150 hours of work on the maps as well. The notebooks document that she worked on numbers 87, 88, and 90, and Randel tallied her hours, estimating her fee at $205.87. Even with her help, he had to request an extension, in part because of surveying demands upstate. The Common Council gave him until May 1820, "as this is a work of considerable magnitude and importance, one which is to become a public record for posterity and which will not only be an ornamental document but of great public utility." And it noted that the "work thus far completed reflect much credit on Mr. Randel as an artist and correct delineator."[130]

Even so, Randel did not quite meet the extended deadline. In one of his notebooks, ever careful in his records, he lists the delivery dates. The list shows what a stunning amount of work the maps entailed. Between February 1 and December 1, 1819, he completed sixty-three maps; between July 8 and September 1, 1820, he finished the remaining twenty-nine. The urgency for Randel was clear: he needed to be paid. The urgency for the Common Council, less so. But as reluctant as the aldermen may have been to honor the old contract, they had made a wise decision. The farm maps were to become the reference document for many other New York City cartographers. Their power and beauty have endured for two centuries. In his famous six-volume work, *The Iconography of Manhattan Island*, architect I. N. Phelps Stokes referred to Randel's ninety-two maps as "the most complete and valuable topographical record of the period that exists. It is, in fact, the only exact early topographical map of the island."[131]

I Have Promised the Public a Map of New York Island

In the midst of his work on the farm maps, Randel found himself embroiled in another conflict, this time with the Albany city council. During the William Bridges dispute, he had used the newspaper to publicize his work and build support for his claim, taking his case directly to the public. A few years later, when bitter about a professional situation gone awry in Albany, he turned instead to the courts, initiating what appears to have been his first lawsuit as plaintiff. The origins of the conflict can be discerned as early as 1815, when Randel wrote to Teunis Van Vechten, chairman of the committee of surveys, to say he was leaving Albany for New York to get the rest of his instruments and he would return "without unnecessary delay" to commence the survey. He asked that monuments be prepared before his return: "Will you have the goodness to have them in readiness that no unnecessary delay may be occasioned in the work?" The language about "unnecessary delay" was prescient, for the committee terminated the contract in 1816, claiming the work had been delayed.[132]

Randel was infuriated and desperate. He devoted pages in one of his field books to tallies of hours he had worked on the survey, the exchanges and understandings that had passed between him and the committee, the expenses he had incurred (assistants, travel, repairing and constructing instruments). He claimed he had invented instruments specifically for the Albany job and had spent time ascertaining how much they were affected by temperature. He had also spent time "in constructing tables of tangents to the Earth's Surface—to be used with the leveling instrument" and time "constructing tables of the versed sines to be used with the measuring instruments in reducing surface measures to horizontal measures." In sum, he was out $6,461.47 (about $97,600 in today's values) by his reckoning.[133]

By 1819 Randel had determined that the Albany government was not going to honor the contract of its surveying committee—exactly the opposite outcome he had received from the Common Council in New York concerning the farm maps. In July he drafted a long letter to the corporation replete with evidence and witnesses (including Simeon DeWitt, his father, and William), figures, and expressions of dismay. In October the case went before Judge Van Ness in the New York Supreme Court of Judicature. Randel found no joy. The case quickly became about whether the men on the surveying committee could be held personally responsible for Randel's claims, as he was suing them, not the city itself. Although Randel ultimately lost, the lawsuit did not dissuade him from seeking justice in the legal system many more times. He revealed a particular talent at crafting arguments, amassing evidence, compiling facts, tallying costs—very much in keeping with his mathematical mind, with his obsessive bent. The Albany case showed that Randel was not afraid to take on his bosses and future employers—that if he felt wronged, he would not rest until he had defended his view of the situation. Randel had begun to earn his reputation as litigious.[134]

Losing the lawsuit further jeopardized Randel's financial situation. Despite the promise of payment from the Common Council when he finished the farm maps, money was evidently tight during this period—and not just for Randel. The country was in the throes of the Panic of 1819; businesses failed, poverty surged, and debtors filled prisons. There was not much surveying work from the city; Randel seems to have been given only one substantial assignment, to survey the Brooklyn Navy Yard and to set monumental stones there. So he energetically returned to the idea of publishing and selling an engraved map of the Commissioners' Plan. The population of New York City reached 123,706 in 1820, making the city the largest in the United States. Randel may have calculated that there were ever more New Yorkers who might purchase a map and ensure his legacy.

He had revised his idea for the New York City map after 1814, deciding to incorporate not just the areas immediately adjacent to Manhattan, but the entire Northeast. "There are at present no Maps published connecting the States on a large scale," he wrote to Simeon DeWitt in 1815. "I have promised the public a Map of New York Island and therefore cannot get off from publishing one as I would give up all plans of that kind. I must therefore do my best to interest as many persons as possible in this one." Randel was concerned that his map would interfere or overlap with DeWitt's New York State map and was careful to ask for permission to publish his own—and even to offer DeWitt involvement if he wanted it. DeWitt seems to have granted his permission but does not seem to have been involved in drafting, producing, or selling the map.[135]

Living in New Jersey was useful for this second mapping endeavor, because the engraver Randel had selected was Peter Maverick, the man who had done William Bridges's 1811 map and perhaps the most famous engraver of the period. Maverick, who was based at times in New York City, worked in Newark between 1809 and 1820. For several years he was assisted by Asher B. Durand, who was to become a famous artist of the Hudson River School, and it is possible Randel dealt with both men. Work on the map stretched out over several years, with both Maverick and Randel turning to other things in between their work on the map. In 1819, however, Randel began pushing for publication, and he wrote to Maverick frequently. Several draft letters reveal Randel's anxiety about the correctness of the maps. He was concerned that the minutes of the latitude and longitude lines, which form the yellow borders, would be incorrect if Maverick didn't divide the paper evenly. He worried about hatching in the hills, saying, "Before you etch in any of the hills, please send me an impression that I may lay down such hills as I have omitted." He made corrections to place names. He fretted that Maverick's press wouldn't take the paper he had selected.[136]

The map, finally published in late 1820, was unusual—and the converse of the farm maps. The farm maps were on a large scale, as noted, and captured the topography and landscape details of Manhattan Island. The New York City map was devoted to grid and city planning, and the perspective chosen was at such a remove that the landscape had no features; the city was flat, filled with details, bustling. Many maps of that era had insets, and most had a cartouche conveying some governmental authority. Randel's has three cartouches but lacks insets. Instead, Randel used the device of two scrolls seeming to unfurl. A map of Philadelphia, rendered at a large scale, rolls open over Brooklyn and Queens. Manhattan is drawn on a large scroll that could unfurl over the entire Northeast, which appears to be the base map. Randel had thought of scrolls when he first wrote to DeWitt around 1815, but he didn't cite his inspiration. At that time he had also considered making each portion of the map separable, saying, "I intend each sheet shall have a margin so as to be mounted if required as separate maps then every person can suit himself subscribing for his own state or all the States and those who do not choose to have large maps in their rooms may have smaller ones."[137]

Randel's type-dense border catalogues population, structures, distances, and roads for both New York and Philadelphia, as well as some details about the "surrounding country" and about his inventive instruments. He explained his technique for getting distances across the rivers. "Much useful and interesting information will be found upon the above map, particularly for the results of the survey by triangles, from Fort Washington to the Light House at Sandy Hook, giving the breadth of the rivers and distances between various points in that space, from which it appears among others, that the distance from Castle Clinton at the Battery to the Light House (going out at the Narrows) is 17 3/10 miles, and on a straight line 16 66/100 miles," he stated in an advertisement that ran in March 1821 in the *New-York Evening Post*. Randel also credited every map he had relied on, and every surveyor (including William). It was his version of transparency.[138]

Randel had high hopes for his map, which he sought to sell for $5 in colored sheets and for $6.50 if mounted on muslin, set on rollers, and varnished. One record indicated that he even sent a copy to President John Quincy Adams. An editor of the *Evening Post* described the map "as decidedly entitled to the preference,

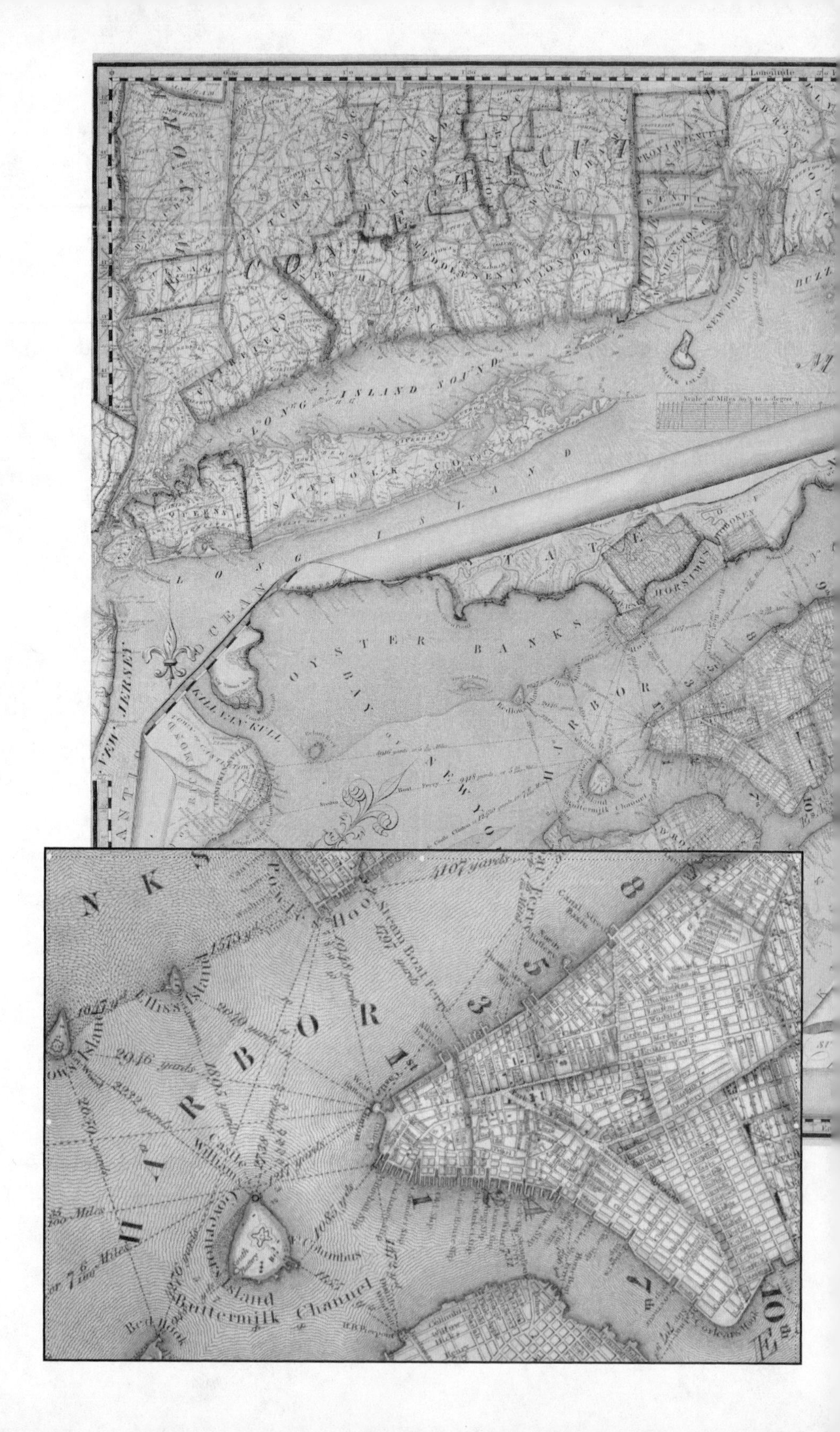

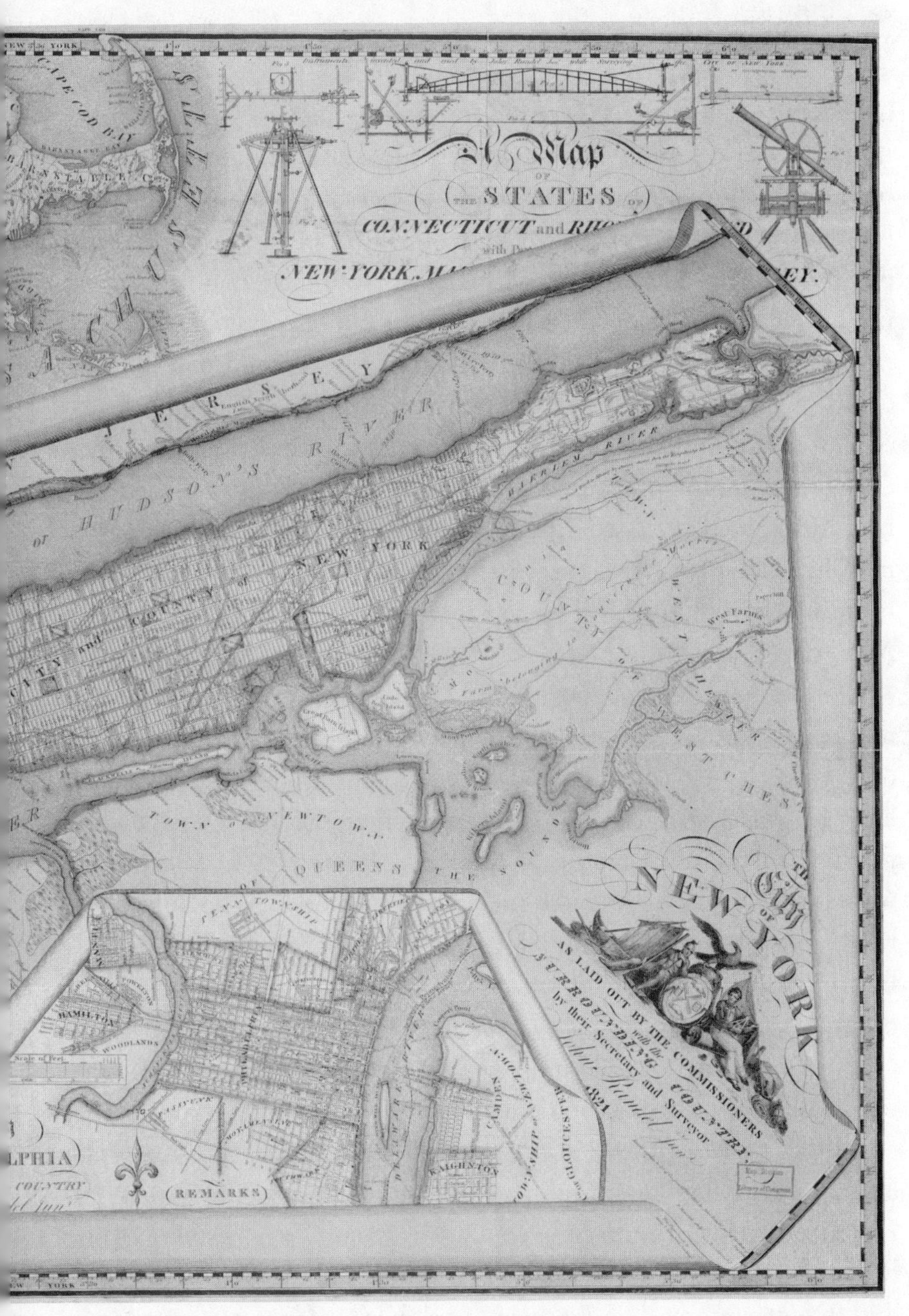

1821 Map of the City of New York as laid out by the Commissioners, with the surrounding country, by John Randel Jr. Courtesy of the Library of Congress.

over any that has ever appeared for extent, accuracy of information and arrangement. We perform a pleasure, therefore, in recommending it in strong terms to public patronage." Beautiful and innovative as it was, the 1821 map "the City of New York as laid out by the Commissioners, with the surrounding country" was not a success. The aldermen noted its art but "are nevertheless of the opinion, that the scale on which it is drawn is too small for the ordinary purposes of reference." The council took twenty-four copies and then turned to David Longworth, author of a New York City almanac, for maps at a greater scale. Randel wrote that he had anticipated scale would be a problem but knew that Bridges's map had been too big to use. He wrote to the mayor saying that he would try to publish a map twice the size of Mr. Longworth's. Apparently he did not.[139]

Randel was keenly disappointed. He had lost a great deal of money paying for Maverick's time, for the plate, paper, and colored ink. He felt the financial setback acutely. Even two years later he was trying to sell the maps. When William wrote complaining of financial troubles, Randel responded that the maps were the best way he knew to raise money quickly: "The only way that I know of unless some job of surveying offers of which I can inform you as soon as that takes place is to again attempt the sale of my maps."[140]

Furthermore, the 1821 map was the culmination of his work on the island. It was to be the enduring, widely disseminated record of Randel's role in the Commissioners' Plan. It had the scope of an important map: it showed New York in geographical context; it included census information. And it was his only engraved map of the city. The 1811 map was done by hand, as were the 1814 map and the farm maps. The promise of an engraved map meant wider sales and perhaps renown as a cartographer. Neither came to Randel. He seems to have made no other significant maps, either in scale or in design, after this one. Today the map is incredibly rare, with only one known colored copy still in pristine condition;

it is housed in the Library of Congress. Another, in a private collection, was destroyed when the World Trade Center fell after the attacks of September 11, 2001.

WITH AN ACCURACY NOT EXCEEDED

Randel's Manhattan era was over. By the end of 1820 he had met all his major obligations. His account for work done between 1811 and 1820 was reviewed by the Common Council: $32,484.98 in total, of which $2,010.77 remained to be paid. Although he would return decades later with a new vision for the city, his next big jobs took him back upstate, to other states, and into different lines of work. His accomplishments on the island had been monumental. He had mapped a vision for an urban future on a scale no other American city ventured, before or since. He had rooted that vision in the landscape with an obsessive care, thoroughness, and exactitude that endure today. He had pleased his patron. Simeon DeWitt wrote that Randel's work on the island, "*was done with an accuracy not exceeded, I am confident, by any work of the kind in America.*" The maps Randel created reveal the features of the island and the landscape as it was more than two hundred years ago. The very act of urban planning ensured a kind of landscape preservation. He had contributed innovations of equipment, technique, and cartographic design to American society.[141]

Randel had also lost a great deal. One important patron, Gouverneur Morris, was dead, as of late 1816. Randel's sterling reputation for accuracy and mathematical genius had begun to acquire the patina of personality. He could be just as exacting and unrelenting about details of finances, contracts, and reputation. In short, he was a character, one who might alienate as easily as not, one who might turn for redress to the press or the courts.

In the coming decades, Randel became increasingly entangled in litigation and financial troubles. He experienced great loss and estrangement. He pursued visionary schemes, made daring suggestions, and risked everything to promote his inventions and his view of projects. Although his reputation for genius grew, he ran afoul of colleagues and companies and came to be described by some contemporaries as erratic and peculiar. He may have been all of those things. He was also ahead of his time.

III

In Which Rose-Redwood Surveys the 1811 Grid and Morrison Surveys Today's

"Now, if you see a pair of sunglasses in here, they are mine. I lost them about two years ago," Morrison calls over his shoulder as he makes his way off a paved path and pushes into a tangle of shrub and trees. "I fell and the glasses went flying and I couldn't find them."

It is September 2008, and Morrison and Rose-Redwood are back in Central Park looking for Randel's bolts. The tree canopy blocks most of the GPS satellite signals and so Morrison is counting off distance in strides—short ones, because of the thick vegetation. Just up the hill from where he is searching, he and Rose-Redwood found a surveyor's mark etched into a flat rock. Morrison thinks it denotes an offset, a line running perpendicular to an established line and used when the established line ran through a house or other large structure. Using a series of offsets, surveyors could set up a short line running parallel to their principal one but circumventing the obstacle. The mark—an incomplete cross and a dot—does not fall in one of the intersections included in the original grid. Morrison tries to get closer to a spot where a bolt might have survived for two centuries.

"Okay, I have paced eighteen, and I have thirty-eight paces to one hundred feet, so that puts me in the ballpark," he says from the thicket.

"But you also said it was twenty-five feet over this way, right?" Rose-Redwood points out.

"We just don't know our line for sure, where straight south is," responds Morrison.

There is silence as Morrison and Rose-Redwood consider their disorientation. Then the metal detector goes wild, screeching and clicking.

"This really is kind of a treasure hunt!" Rose-Redwood exclaims.

Morrison digs a bit. Nothing. The rock-cum-bolt might be too far below the surface to get at with a quick, shallow dig. More likely the detector picked up a bottle cap or a foil gum wrapper, or the iron of the Manhattan schist.

After several more minutes in which the invisible intersection doesn't reveal itself, the GPS receiver is turned off. The tripod is collapsed. The surveying party moves north; Morrison and Rose-Redwood catch a bus uptown to 120th Street and Marcus Garvey Park.

It is a warm, clear fall day, but the summit of Marcus Garvey Park is empty. Salsa music wends its way up the hill and through the trees from the street below. Morrison and Rose-Redwood, digesting hastily consumed pizza slices, start hiking over rocks, avoiding the many shards of glass, while the GPS receiver starts recording and calculating its coordinates. They search along the line of Fifth Avenue. On an earlier, solo outing Morrison had found a bolt there, but it was not in the right spot, and he thinks it was set during a different survey. Again, after an hour examining the intersection where Randel noted he set a bolt and where he recorded an elevation of 96.29 feet, no luck.

Obsession being what it is, though, Morrison and Rose-Redwood will be back.

Rose-Redwood discovered Randel in *Manhattan in Maps*. Augustyn and Cohen's beautiful book, published in 1997, was a revelation for many readers. The depictions of New York City between 1527 and 1995 celebrated little-known cartographers and views and captured the island's early landscape as well as the city's march up the island from the harbor to Spuyten Duyvil, where the Hudson and Harlem Rivers meet. The book included Randel's 1811 Commissioners' Plan, his 1814 version, his 1821 map of the city and surrounding country, and farm map number 27. It was farm map 27 that captivated Rose-Redwood.

In the fall of 2000, Rose-Redwood had just begun his graduate work in geography at the University of Pennsylvania, and he wanted to study the environmental history of New York City. His thoughts about nature and the city first turned, as most people's do, to Central Park. But they didn't settle there. "I began to realize that the grid was, at least in terms of the landscape, one of the major transformations of New York in the modern period, at least since the Europeans arrived," he says. "When it comes to the complete reconfiguration of all the topography and the spatial structure of the island, it was the grid that really shaped that." He began reading about the history of the grid and looking for materials describing the island pre-grid, which took him to farm map 27. The map, which Randel delivered to the Common Council on September 20, 1819, illustrates Turtle Bay, the area between 45th and 53rd Streets and between Third Avenue and the East River. The Eastern Post Road—the famous 270-mile Colonial road to Boston—travels diagonally across the page, passing the properties of James Beekman, Stephen N. Bayard, the heirs of Francis Winthrop, and others. Randel had captured wonderful landscape details. Rose-Redwood loved the ragged, rocky shoreline, hills, a small cliff, and a creek. He loved this valuable view of the pre-grid landscape—and there were ninety-one more, covering the entire island.

Rose-Redwood soon traveled to visit Randel's maps. In the spring of 2001, he spent many days at the Manhattan borough president's office in the dimly lit windowless map room, examining four bound volumes, the whole map readable as a series of books. (Today the maps, fully restored between 2005 and 2006, are kept unbound.) Each individual map is big, about 3 feet by 2 feet, and one must climb a rickety metal footstool and lean over a long, high metal table to look closely. Rose-Redwood was first helped by Brian Cook and then by Hector Rivera, who cares for and consults the 3,300 legal and 1,000 reference maps housed in three short aisles of metal cabinets and wooden map drawers. Randel's maps are in good company, with a rare map from 1749, late eighteenth-century maps by Casimir Goerck, and map clerk Otto Sackerdorff's 1872 maps of the island north of 155th Street.

As he peered at the maps, Rose-Redwood discovered small, lightly inked notations, "mon" or "bolt," and a series of numbers at intersections along five of the avenues. Randel's notations showed the elevation of the land at those spots: a monument at what was to become Fifth Avenue and 59th Street sat at 29.99 feet of elevation. Rose-Redwood decided to enter those values, from intersections around the city, in a Graphical Information System program, or GIS model, and produce a visual depiction of the topography at those intersections. He wanted to help New Yorkers recover a lost landscape and to "thereby rejuvenate the historical consciousness of an ever-changing present." New Yorkers have always been curious about recapturing the island's past, he noted, even when the grid was not so old. As historians in the late nineteenth century put it: "We know the island is no longer young; that the laughing streams and smiling valleys which once dimpled its face have given place to the countless hard wrinkles, called streets; but, in the heart of each faithful old lover of the land of the Manhatoes, the memory of its rippling waters is still an everflowing spring of pleasure."[1]

A few months later, Rose-Redwood—with his guitar, laptop,

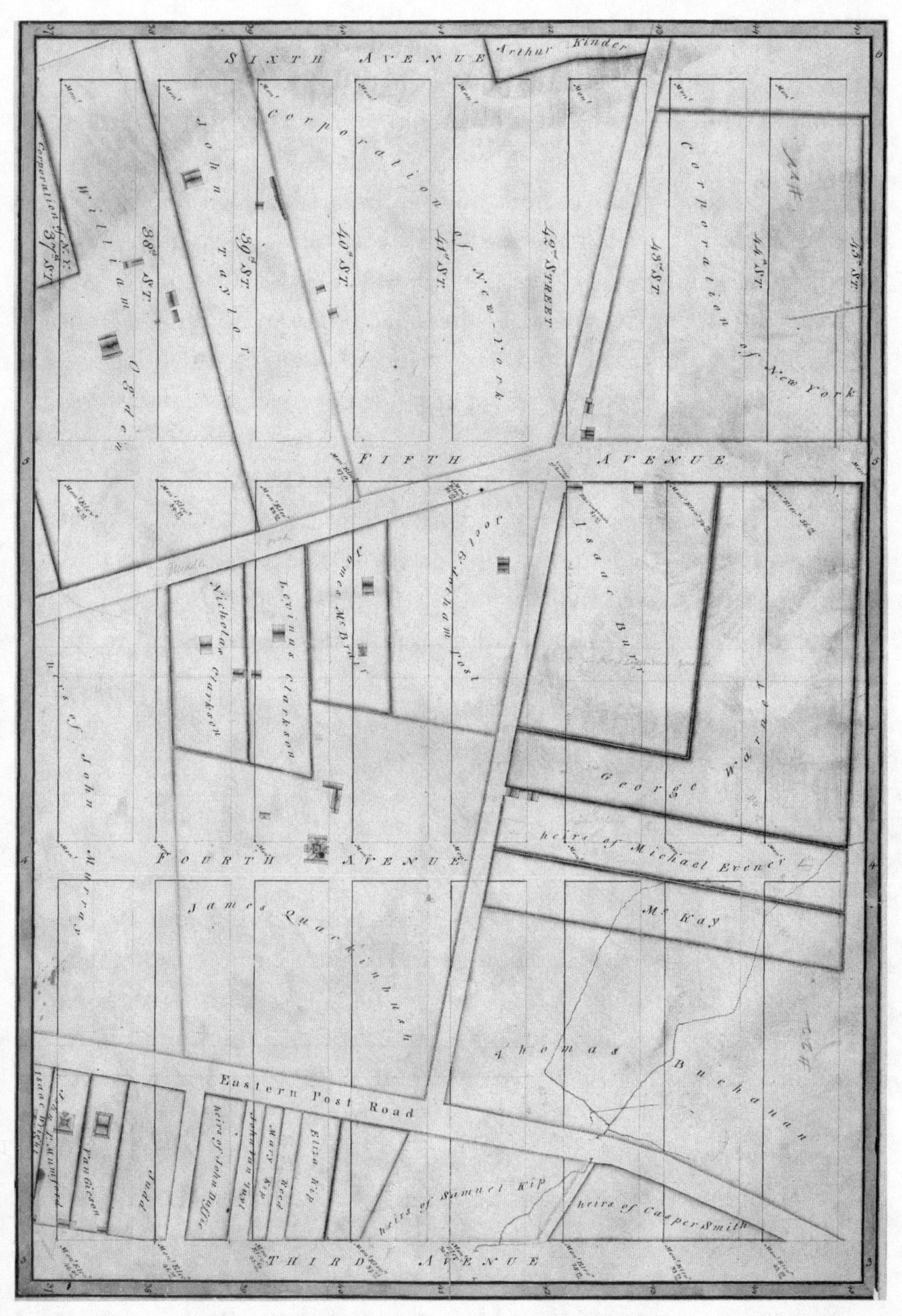

Farm Map No. 22: some houses can be seen in the middle of streets-to-come. Courtesy of the Manhattan borough president's office.

and many books—arrived to spend the summer in New York. He moved in with his grandmother, Lillian Rosen, won an Environmental Protection Agency fellowship, and set about daily treks to the New-York Historical Society, Columbia University's library, the municipal archives, or the Manhattan borough president's office. As he traveled the grid—the 86th Street crosstown bus, the 1 train up Broadway or the 6 down Lexington Avenue to the foot of the Brooklyn Bridge—he read philosophy: Michel Foucault, Bruno Latour, Max Weber, Henri Lefebvre, and Friedrich Nietzsche. As sometimes happens during an obsession, everything he read, everything he saw, seemed to relate to the grid; even a viewing of *Planet of the Apes* resonated because it showed a New York icon as it might look in the future. "The grid is what everything has in common through the spectacles of the western mind," he wrote in his journal. "It is, indeed, the projection of that very mind upon the world, thereby transforming the world into the image of the gridded gaze."

One Map's Datum Is Another Map's Doom

It took Rose-Redwood two days to record every elevation from the farm maps that he had not recorded on his earlier visits; there were 600 elevations in total; fewer than half the intersections had them. He was also missing a crucial piece of information. At the intersection of Eighth Avenue and 65th Street, for instance, Randel recorded the elevation as 59.06 feet. But in relation to what? If Rose-Redwood entered the elevations into a computer program and recreated the topography, he would get peaks and valleys only in relation to one another. The figures would have no meaning in terms of Manhattan's current landscape unless he was able to correlate the old data with today's. Rose-Redwood had hit the datum.

"Datum" is the term used by engineers, surveyors, navigators, GPS experts—by anyone involved in interpreting or traveling landscapes—for a reference system that locates points in space. A datum anchors information about position by grounding a surveyor's measurements in relation to the earth's physical spheroidal reality and by providing a common coordinate system. Datums are based on bench marks, which serve as the referents to which other measurements are compared: how much higher or lower is something than a certain bench mark. A vertical datum, as one would expect, describes elevation and is typically based on sea level. The U.S. version is called the North American Vertical Datum of 1988, or NAVD 88. It replaced the National Geodetic Vertical Datum of 1929, which averaged information from twenty-six sea-level monitoring stations. Over decades, regional variations in sea level and shifting land-based bench marks undermined the 1929 datum. Experts at the National Geodetic Survey—the government agency that concerns itself with location, location, location—established the NAVD 88 using satellite information. The satellite data are more comprehensive and can correct for variations in sea level due to, for example, gravitational pull.

Horizontal datums provide latitude and longitude, or coordinates—how far north or south or east or west, as opposed to how high, how low. The current horizontal datum is called the North American Datum of 1983, or NAD 83. It was preceded by two others, those of 1879 (which came to be called the 1901 datum) and 1927, which were based on then current scientific understanding of the shape of the earth and on surveying markers set in the nineteenth and early twentieth centuries. The origin point for the 1927 datum, a metal disk embedded in a concrete slab, was fixed on a ranch in Kansas, near the geographic center of the country. With the advent of GPS, the datum went global. Whereas one could visit and touch the 1927 datum, the North American Datum now resides very close to the center of the planet. "Unless you happen to be

Jules Verne, you can't get there," says David R. Doyle of the National Geodetic Survey, who is described by some surveyors as the "godfather of geodetic surveying in the United States." The NAD 83 is linked to an international model of the earth: the Geodetic Reference System 1980, or GRS 80. That model, based on mathematical calculations, is called the reference ellipsoid and serves as the best contemporary estimate of the true figure of the earth.

National and global datums have evolved as models of the earth changed, as understandings of sea level and regional variations thereof changed, as technologies changed—a process called datum shift. The same is true of regional and local bench marks, and many cities have a series of datums. Chicago, for instance, used the average level of Lake Michigan instead of average sea level, the sea being impractically far away. Elevations were thus higher or lower than the lake, which was the zero value. This Chicago datum can now be related to the national sea-level-based datum if needed: in 1929 it was 579.48 feet above the national zero value.[2]

New York City is dense with datums, dizzyingly so. Seven vertical datums and at least thirteen coordinate, or horizontal, datum systems existed even before state and national datums were introduced, according to Scott S. Zelenak, a surveyor with the Port Authority of New York and New Jersey and an expert on the region's datums. So when engineers or others describe a place, they must specify which datum they are using. Making one datum correspond to another can be straightforward, as in the case of Lake Michigan. But such correspondence can be mind-bendingly complicated when a city's history resides in many systems, maps, and measurements, all using different datums and all made in different ways with different instruments or techniques. Doyle, who strives with Randelesque persistence for datum perfection and unification, says he wishes he kept a bottle of single malt in his desk for the times when people call about New York City. "I can hear that whimper in their voice when they say, 'I am working in

New York City and . . . ,'" he says. After the September 11 attacks, datum profusion posed problems for the Federal Emergency Management Agency and other agencies, he explains. "You would hear things like, 'If I have heights in this coordinate system, do I add 1.3 feet or subtract 1.9 to get it to correspond to my GPS coordinates?'" Doyle says his mentor worked for decades to devise algorithms that would convert every New York City datum into every other—and into the national datums. It still hasn't happened.[3]

It was essential that Rose-Redwood discover Randel's datum. And by midsummer he had. On a Thursday in late July, he traveled to the farm map room and met with the city's then chief topographical engineer, Eric Cardwell, to compare Randel's elevations with a current elevation map—one referenced to the vertical datum of 1929 (and convertible to the 1988 datum if needed)—to see how much they might differ. Rose-Redwood also took to that meeting a list of sixteen elevations Randel had included on his 1811 map of the Commissioners' Plan. Seven were for intersections where elevation data were also recorded on the farm maps. Of those, only four were in the right intersections and overlapped exactly: the same corner of the same intersection. Strangely, the 1811 numbers differed from the corresponding farm map numbers. Rose-Redwood was flummoxed. He and Cardwell huddled together for more than an hour, comparing the numbers.

As Rose-Redwood and Cardwell looked at the values, it became clear that Randel's original datum corresponded to a Manhattan datum still in use. The 1811 map elevations were based on a mark above the high-tide line—a mark Randel had made at the base of a column at the southwest corner of the old almshouse at Bellevue Hospital, and which he had then also marked at City Hall. Rose-Redwood later discovered an article on just this issue. According to city engineer Frederick W. Koop, who described the locations of the tide markers in *Precise Leveling in New York City*, Randel's Bellevue measurement is the oldest recorded elevation bench mark for the city and came to be referred to as the Public Works Datum or the City Datum. That 1811 bench mark is 2.75 feet above the National Geodetic Vertical Datum of 1929.[4]

For reasons unknown, Randel did not use that same bench mark on the farm maps. Instead he used an average between high and low tide. But because Rose-Redwood could compare four elevations that appeared in exactly the same spots on the 1811 map and the farm maps, he could design a conversion. For instance, at the northeast corner of the intersection of Third Avenue and 34th Street, Randel gives an elevation of 34.03 feet on the farm maps and 28.33 feet on the 1811 map. The difference is 5.70 feet. Making the same calculations for two other intersections in common on the two maps, Rose-Redwood came up with an average difference of 5.63 feet (he opted to disregard one wild outlier). That meant Randel's farm map datum was 5.63 feet below his 1811 datum, or 2.88 feet below the 1929 datum.

It may seem like a small adjustment in calculations, but Rose-Redwood now had the data he needed to compare Manhattan as it was in the early nineteenth century to Manhattan today. Which meant he could bring the original topography to life in terms that any New Yorker could relate to. It also meant that Randel's legacy extended beyond the grid, which was the city's first horizontal coordinate system and is called the "legacy datum." Every official elevation taken in Manhattan since 1811—thousands, likely millions of them—relied on Randel's foundational work.

Getting into the Grid

Despite his datum discovery, Rose-Redwood says his advisers encouraged him to do a more scholarly assessment of the grid instead of a GIS computer model, so he began to characterize and list the houses and structures noted on the maps above

North Street—all 1,865 of them. The results were astonishing: ash houses, a bathhouse, a boat house, churches, dairies, dwelling houses, henhouses, ice houses, forts, root cellars, schools, and stables. About 40 percent—721 structures—sat smack in the middle of streets and avenues that would soon materialize.

It was no surprise property owners had set their dogs on Randel, no surprise he had been repeatedly arrested. When the young surveyor arrived on their land, these nineteenth-century New Yorkers must have known their buildings were doomed. Those early angry reactions foreshadowed the full-blown grid antipathy that was to emerge slowly over several decades. The best-known early criticism was Moore's *A Plain Statement Addressed to the Proprietors of Real Estate, in the City and County of New-York*—the Moore who had turned to Randel for topographical support when he lambasted city street-graders for their insensitivity to elevation and water flow in Chelsea. Moore didn't attack the grid per se, rather the leveling that came to accompany its implementation. "The great principle which appears to govern these plans is," he said, "to reduce the surface of the earth as nearly as possible to a dead level. The natural inequalities of the ground are destroyed, and the existing water-courses disregarded."[5]

Most New Yorkers did not initially critique the plan once it was unveiled. "In 1811 the gridiron had been so widely accepted as the optimal street arrangement for a commercial city that the plan received only perfunctory treatment in the press—even though it had a dramatic impact on existing property lines," writes David Schuyler in *The New Urban Landscape*. Some even saw the plan as a welcome antidote to the chaos that pervaded the old streets. In his writing about the grid, legal historian Hendrik Hartog cites a positive assessment written in 1814: "The arrangement of the original or lower part of the city . . . is essentially defective. Beauty, order and convenience seem to have been little valued by our ancestors." Many residents also came to embrace the ability to plan for the future. By 1836 positive opinion prevailed, as Hartog shows in this quote from a city official: the Commissioners' Plan "laid out the highways on the island upon so magnificent a scale, and with so bold a hand, and with such prophetic views, in respect to the future growth and extension of the city, that it will form an everlasting monument of the stability and wisdom of the measure."[6]

As the grid spread up the island, however, it engendered more criticism. Over the past two centuries it has been censured on many fronts. The streets are too narrow, producing congestion. The blocks are subdivided in such a way that building lots are long and thin (generally 25 by 100 feet), depriving residents in the center of blocks or at the backs of buildings of air and light. No provisions were made for rear alleys, as in the Philadelphia grid plan, which predates New York City's by 125 years. Monumental or modern architecture had no room to flourish, until superblocks such as Rockefeller Center and Lincoln Center were created—occasionally giving rise to even more congestion. The grid had no aesthetic, no beauty. Edith Wharton denounced New York's grid-bound lack of character, calling it "this cramped horizontal gridiron of a town without towers, porticoes, fountains or perspectives, hide-bound in its deadly uniformity of mean ugliness."[7]

Many other writers have concurred. Of the city's natural spots, Edgar Allan Poe wrote, "In fact, these magnificent places are doomed. The spirit of Improvement has withered them with its acrid breath. Streets are already 'mapped' through them, and they are no longer suburban residences, but 'town-lots.'" In the 1840s Poe swam in Turtle Bay and sculled around Blackwell's Island, now Roosevelt Island. Poe lived in the area portrayed in Randel's farm map 27, the area that had fascinated Rose-Redwood.[8]

Perhaps the most passionate mid- to late nineteenth-century objections to the grid related, as Poe's did, to nature. Moore had rued the razing of hills in 1818, and as the century wore on, he had company. "I always think it a pity that greater favor is not given to the natural hills and slopes of the ground on the upper part of Manhattan Island. Our perpetual dead flat, and streets cutting each other at right angles, are certainly the last things in the world consistent with beauty of situation," Walt Whitman wrote in 1849. New Yorkers increasingly understood the implications of the grid plan. Schuyler writes, "As construction swept northward on Manhattan Island, destroying pastoral farms and densely wooded hills, the rationalism of the gridiron began to seem more and more irrational. Increasingly those spokesmen dissatisfied with the gridiron criticized it for depriving residents of daily contact with nature."[9]

Although the Commissioners' Plan of 1811 had no curves, it had included parks and public squares. But they were small and few—perhaps 500 acres in total. The commissioners stated that because the island was embraced by "those large arms of the sea" and was not on the "side of a small stream such as the Seine or the Thames," it did not require many vacant spaces "for the benefit of fresh air and consequent preservation of health." The populace of the burgeoning metropolis did not agree. As the city grew—by 1830 it had reached what is now Union Square at 14th Street; by the mid-1850s, 42nd Street; by the 1870s, 125th Street on the East Side and 59th Street on the West—some designated open spaces disappeared (the Parade) or shrank (Market Place and Union Place), and residents increasingly called for parks. In the 1830s land donation or legislation gave rise to Gramercy Park, Tompkins Park, Stuyvesant Square, and Madison Square. And in 1844 William Cullen Bryant, the editor of the *New-York Evening Post*, initiated the campaign that led, ultimately, to Frederick Law Olmsted and Calvert Vaux's 1858 Greensward Plan, the nineteenth century's greatest constructive criticism of the grid: Central Park.[10]

The linearity advocated by the commissioners and political leaders rooted in the Enlightenment of the early nineteenth century was being rejected by those of the ascendant Romantic movement. Olmsted, who was appointed architect in chief of the park and who has since been the cited spokesman for its ideology, was of a Romantic sensibility. He believed that nature elevated men's souls and inculcated good morals. Olmsted intended both

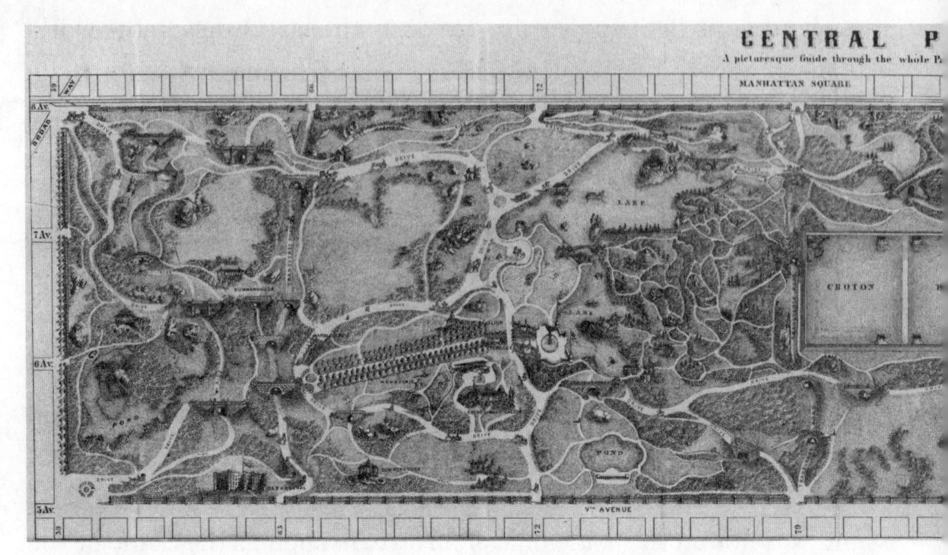

1865 map of Central Park by Prang & Co. Courtesy of the David Rumsey Map Collection, www.davidrumsey.com.

the pastoral landscape of Central Park and the rougher northern portions to generate a "sense of the superabundant creative power, infinite resource and liberality of Nature." He derided the grid, its lines and its levels, as Spiro Kostof, author of *The City Shaped*, notes: "Curved streets 'suggest and imply leisure, contemplativeness, and happy tranquillity,' Olmsted wrote, in contrast to straight streets which implied 'eagerness to press forward, without looking to the right or left.'"[11]

As was true of some of his contemporaries, Olmsted was not interested in untended nature, and Central Park is neither wild nor natural in the way many people define "natural" today. "Olmsted designed parks to mediate between the intense artificiality of the city and the harshness of nature in its wilderness state. He did not want to simulate the awesome grandeur of a forest within an urban park, nor did he want large areas of ornamental gardens which evidenced the fussy intervention of man's art. Rather, his intention was to re-create within the city the scenic qualities that had once been easily accessible beyond the fringe

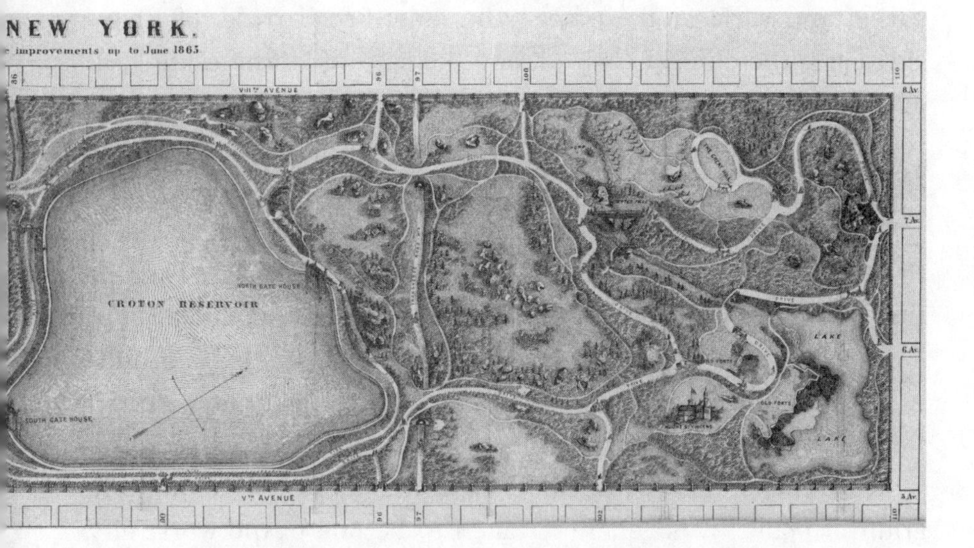

of urban settlements. He wanted a country road, not a boulevard or a wilderness trail," writes Leonard J. Simutis. The park is a masterpiece of landscaping and stage-setting with living props, a masterpiece of engineering—roads, drainage, rocks sculpted with gunpowder, sized to suit the scene.[12]

But for all its constructed, calculated, gardened qualities, the park preserves some elemental topography. Manhattan's primordial terrain persists in places, in rocks, and in the scars striating the schist—the cuts left by glaciers flowing across the land as recently as 15,000 years ago during the Wisconsin Glacial Episode. "The time will come when New York will be built up, when all the grading and filling will be done, and when the picturesquely-varied, rocky formations of the Island will have been converted into formations for rows of monotonous straight streets, and piles of erect buildings. There will be no suggestion left of its present varied surface, with the single exception of the few acres contained in the Park," Olmsted wrote with foresight in 1858. Those 843 acres are not the only remnant of the island's varied and variegated surface—the long, broad rocks of Riverside Park, the cliffs of Inwood, and the spur of Morningside Park all provide glimpses. But Central Park and its outcroppings and rocks are for many New Yorkers the principal connection to the island as it was before the city.[13]

In many people's eyes, however, Central Park did not redeem the grid that dead-ends against its low-lying stone walls. "The fact that it was this gridiron New York that served as a model for later cities was a disaster whose consequences have barely been mitigated by more modern city planners," writes historian John W. Reps in *The Making of Urban America*. And in his essay, "The Grid as City Plan," Peter Marcuse contends that "Manhattan's gridiron plan . . . is generally taken to be one of the worst city plans of any major city in the developed countries of the world."[14]

The Grid's Been Everywhere

Meager primary source material explains why an undulating, corrugated island was given orthogonal form. By necessity, grid scholars turn to the commissioners' published remarks of 1811—which accompanied William Bridges's version of the map, not Randel's—to understand the choice. Letters between members of the Common Council and the commissioners, or between the commissioners themselves, discussing the genesis of the grid or alternatives have not emerged, if they existed.

Documentary lacunae have not prevented, and perhaps have actively encouraged, the creation of origin myths. Rebecca Read Shanor, who studied the grid plan and wrote *New York's Paper Streets* and *The City That Never Was*, notes that Olmsted claimed a mason's sieve placed on a map led the commissioners to their choice: "What do you want better than that?" they apparently asked. Another story holds that the commissioners caught sight of a shadow cast by a gravel screen and exclaimed, "There you are! . . . That's my idea exactly." Other historians credit Simeon DeWitt's interest in classical antiquity and the grid plan of Rome.[15]

For their part, the commissioners wrote simply that the grid appeared to be the best plan, "or, in other and more proper terms, attended with the least inconvenience." They explained away embellishments, "those supposed improvements by circles, ovals, and stars"—which some historians note refer deridingly to Washington, D.C., at that time a muddy, depopulated expanse—as not useful for the habitations of men. They cited economy: "Straight-sided and right-angled houses are the most cheap to build and the most convenient to live in." They cited continuity: "An important consideration was so to amalgamate it with the plans already adopted by individuals." Yet the commissioners' conceptual remarks were pithy. The major portion of their document reviewed particulars about streets at a fine-grained, Randel-informed level. One example among dozens: "The northern side of Forty-second street touches the southern side of Monuments Nos. 1 and 42, placed four-tenths of a foot eastward of the westerly side of the First avenue."[16]

Perhaps logic-of-design remarks were slight because there was no need to justify a grid: it was the a priori blueprint. After all, Casimir Goerck had suggested essentially the same plan just a few years before the commissioners were appointed. City-on-a-grid was nothing new in America. Philadelphia had a grid, as noted. Savannah did too, as did other early towns. Cities had been built on grids for centuries, in Italy and Greece, in Mexico, Central America, Mesopotamia, China, Japan. The grid is the most pervasive city design on earth. And the grid as a land-planning tool, not just as city plan, was sweeping the United States at that time because of the 1785 land ordinance. The grid seems to have been unremarkable to most contemporaries.[17]

The grid certainly resonated with the political values of the young country. Hartog argues that the grid was presented "as the antithesis of a utopian or futuristic plan." The commissioners emphasized "ordinary" ways of life, and made clear that "government ought not to act in such a way as to create inequality or special privilege. Again, only a uniform grid could possibly meet this standard," he writes. "The reconstruction of the natural environment to fit the requirements of republican authority might be termed the hidden agenda that underlay the commissioners' seemingly modest proposal . . . The commissioners wrote as if all they cared about was protecting the investments of land developers and maintaining government-on-the-cheap. The plan that resulted, however, served to transform space into an expression of public philosophy." That philosophy espoused equality and uniformity. "In a city shaped by rectangular blocks, all structures and activities would look roughly the same. Individual distinctions, whether cultural, charitable, economic, or whatever, would have to find their place within a fixed, republican spatial organization."[18]

If you break apart a hologram, each piece of shattered glass contains the entire image the intact hologram portrayed. If you travel though a fractal, the pattern is repeated at a different resolution—the larger image comprises itself writ small. New York City had a similar relationship to the larger republican spatial organization of the country it was poised to play such a crucial role in, commercially and culturally. New York was the largest urban application of a grid within a country that had embarked on the world's largest territorial application of a grid. On both scales, the grid embodied, to many political leaders, egalitarianism though uniform geometry.

New Yorkers and Americans came to live that particular geometry at a time when principles of reason were perceived as just as crucial to successful nationhood as were principles of egalitarianism. In her book *A Calculating People*, Patricia Cline Cohen argues that teaching mathematics became more important in the nineteenth century because numeracy became associated with the "fostering of democracy . . . As widespread rational thinking came to be perceived as necessary to the workings of democracy, educators looked to mathematics as the ideal way to prepare a republican citizenry." Reason and logic were necessary requisites for the kind of society America was seeking to create.[19]

Elevating Effects

Mathematics proved a crucial consideration for Simeon DeWitt in planning the grid as well. Analyses of the commissioners' thinking about the plan have been, as mentioned, largely based on the remarks accompanying the 1811 map. But other documents

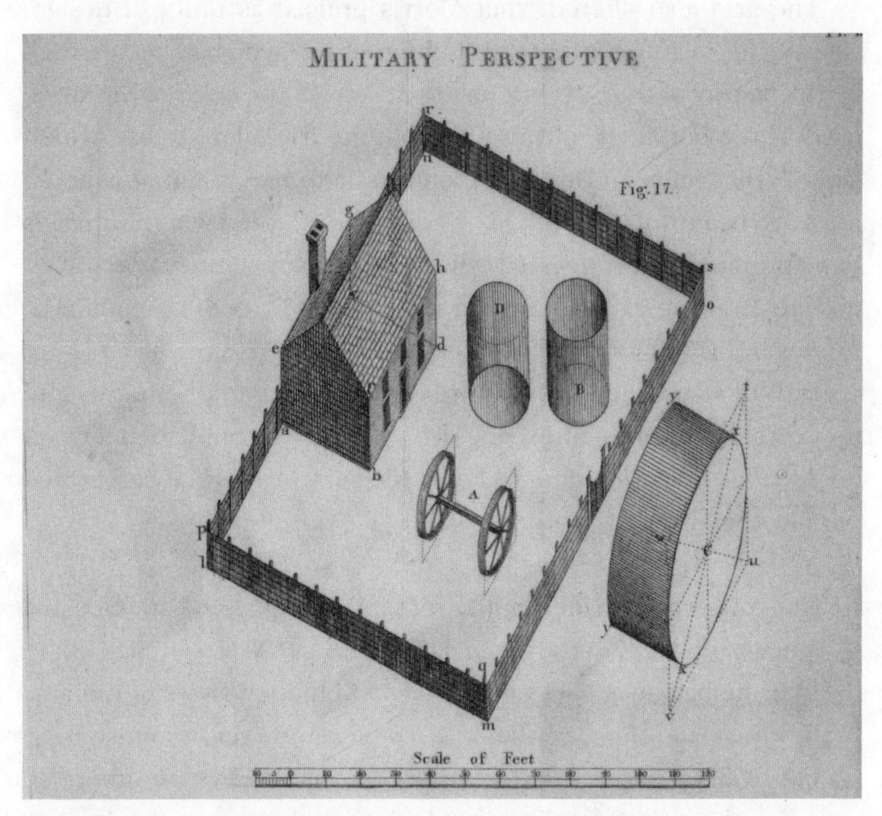

Military Perspective from Simeon DeWitt's *The Elements of Perspective*. Courtesy of the American Antiquarian Society.

by DeWitt and Gouverneur Morris (little survives of Rutherford's writings) suggest that for them the grid also embodied both beauty and spiritual values.

Morris, who had traveled extensively in Europe and was familiar with different city forms, has been credited by some scholars as the leader of the group of three commissioners. Between 1807 and 1811, Morris's name occurs frequently in the *Minutes of the Common Council*, and it was Morris who officially put forth Randel for the task of setting monuments and bolts. Morris had little faith in human rationality, believing that people were easily led astray by "passion, by indolence and even by caprice." The grid plan pushed back on such disorder, mentally and physically, arguing for what Morris called "the boundless Perfections of the Deity."[20]

The grid also offered what Morris praised as unity of design. "During his visit to Bath in the summer of 1795, he had been struck by the beauty of the city, but its appeal was in the color of the stone used in the buildings, not in its romantic, irregular streets, which lacked 'that unity of Design which seems to me essential. Curved lines are beautiful and should certainly be adopted when Nature is to be imitated and concealed,' the future urban planner noted, 'but no one I believe ever heard of a natural City,'" quotes William H. Adams in his biography of Morris.[21]

DeWitt's aesthetic arose more directly from mathematics and its expression in the landscape. In 1813 DeWitt published a small book called *The Elements of Perspective*, in which he celebrated perspective drawing, noting that it

> gives him who is made familiar with its principles and practices a new and deeper interest in THE APPEARANCE of THINGS. By it he becomes habituated to discriminating views of their beauties, and thus they acquire a superior power of ministering to his pleasures. In the aspect of nature, where others see nothing to affect them, but look "with brute, unconscious gaze," he sees the distinct myriads of parts, wonderfully

formed and put together by infinite wisdom to constitute a whole, perfect in all the varieties of proportion, shape, color, and purpose, and his sensations are absorbed and dissolve in the harmony that reigns universally among them, Delight streams into his soul from every quarter to which he turns the contemplative eye.[22]

Rose-Redwood, who had been reading Morris and DeWitt's writings, came to view the grid as a religious and aesthetic choice as well as an economic and egalitarian one. "Rationalizing the landscape would instill into the youthful population the discipline necessary for achieving the worldly goals of obtaining wealth and power," he wrote, but would also bring them closer to God. Science, Rose-Redwood explains, was a route to "the higher regions of bliss" for DeWitt. The science of surveying, in particular, was a way of touching harmony, symmetry, beauty, and perfection.[23]

In *The Elements of Perspective*, DeWitt described two devices he had constructed that would enable people to appreciate geometry and perspective and to be able to draw—a crucial skill to have, in his view. "The productions of the creative mind grow under the pencil till they result in wonderful systems, endowed with powers to produce effects of incalculable benefit to man," he wrote. The devices were called a diorascope and a dioragraph. The first one permitted an individual to look at a scene through a frame divided into a grid by threads or wires and then to copy "each square of the picture on the corresponding square of his paper"—an approach used by artists since the Renaissance. The second was a complicated device with a moving port-pencil, rods, and grooves, which "*draws* the outlines of pictures *mechanically*." Many pages of description do little to clarify how the device worked. But, according to DeWitt, the dioragraph would allow "true perspective drawings of any object be made, with nearly as much facility as copies of pictures are taken, by laying paper on

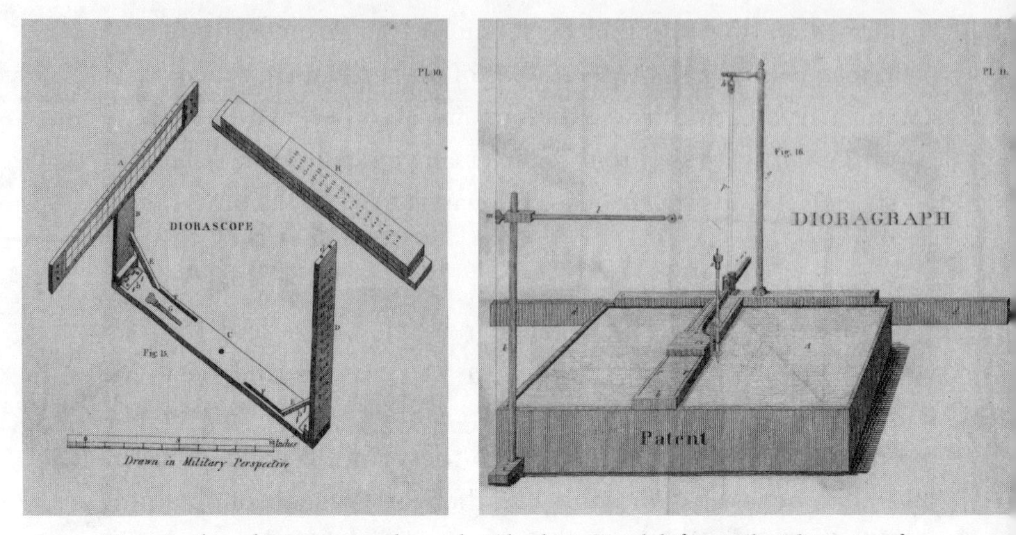

Dioragraph and Diorascope drawn by Abraham Randel, from *The Elements of Perspective*. Courtesy of the American Antiquarian Society.

them and tracing on it the lines of the original, seen through it when held up to the light."[24]

A diorascope would be relatively easy to make, a dioragraph would not. Indeed, wrote DeWitt at the end of his book, dioragraphs could be purchased. They "may be had of Mr. Abraham Randel, Cabinet-Maker, Albany, who has the right of making and vending them." John Randel Jr. became involved as well, for DeWitt had Abraham construct some of the machines and send them to New York City, where Randel served as salesman. "I called yesterday at Vecchio's to hear about the Dioragraph, several persons have seen and admired it, but think that price is too great," Randel wrote to DeWitt a year after *The Elements of Perspective* appeared. "I think if some cheap ones were made without any box or even drawing board, they would sell much better and be much more profitable, at present these do not appear to be sufficient encouragement to send any more such costly ones down, if you send five or six simple instruments down I will distribute them in different stores in the City."[25]

By embracing the grid as a planning form—and by replicating it on a personal scale to cultivate individuals' artistic skills and appreciation of geometry and beauty—DeWitt sought to bring about widespread spiritual transformation through mathematics. Rose-Redwood came to the conclusion that DeWitt was not just organizing the landscape in an easy and popular way, seeking to ensure a democratic and financially secure future; he was also imparting beauty and harmony. Citizens could opt to see their world that way; perhaps they could purchase a dioragraph. But they could come to *live* that harmony too. Their surroundings would inculcate it without people even appreciating their transformation. If "people would not pursue the aesthetics of harmonious proportion willingly, then they must be forced to do so—it must be 'impressed' on them," Rose-Redwood writes.[26]

ISLAND OF HILLS OR CARTESIAN FLATLAND

An animating question for Rose-Redwood was whether the imposition of DeWitt's aesthetics of harmonious proportion had altered the physical island of Manhattan. Had the city used the 1811 plan as the justification or impetus to flatten the island, to level hills as beautiful as those of Rome? What had the Commissioners' Plan really meant for the topography, for the ancient landscape of Manhattan?

Just as Rose-Redwood returned to his GIS elevation model, his thesis complete, he was contacted by Eric Sanderson at the Wildlife Conservation Society, which is based at the Bronx Zoo. Sanderson, a landscape ecologist specializing in GIS, and William T. Bean, then an undergraduate studying ecology at Columbia University, were also building a pre-grid topographical computer model of the island—the foundation of what was to become, six years later, Sanderson's popular book *Mannahatta*. Sanderson invited Rose-Redwood to meet for lunch at the Bronx Zoo, and the two discussed one of their heroes, environmental historian William Cronon, who lifted the myth of America as a wilderness before Europeans arrived by revealing a landscape shaped by indigenous people. With his second book, *Nature's Metropolis*, Cronon brought cities into the realm of environmental history.

The two ended up working separately. Bean and Sanderson did their own research, and Rose-Redwood went on to make two different models of Manhattan's original topography. *Re-Creating the Historical Topography of Manhattan Island* came out in 2004. In that study, Rose-Redwood explains his exploration of Randel's datum and uses the popular GIS computer program ArcView to create a first-cut, three-dimensional model of the island pre-grid. There are some shortcomings: the model has only three reference points to link the farm map datum to the 1811 map datum to the 1929 vertical datum. The data exist only at intersections, so ArcView averaged in-block elevations. Finally, the topography depicted is that of the early nineteenth century and uses the 1929 datum as a reference. In other words, the nineteenth-century elevations can be seen in twentieth-century terms, but they are not yet compared with the actual elevations from today.

As a first step, however, the model was a success, and it would not be unfamiliar to many modern Manhattanites. The East Side appears flat overall, with a dramatic hill at Marcus Garvey Park. The West Side is higher overall and has more hills, with the exception of the area around 125th Street, where a fault cuts eastward and where Hudson and East River waters used to meet and mix during very high tides. The difference between the highest and lowest points is 154 feet—the height of the Green Lantern roller coaster at Great Adventure, roughly the height of the lighthouse at Cape May, New Jersey.

Seven years later, Rose-Redwood elaborated on his initial findings. In 2011 he and Li Li, a former colleague and GIS expert at Texas A&M University, where Rose-Redwood taught for several years, published a new study, more robust than the first. The two used a late twentieth-century digital elevation model, or DEM, and compared today's elevations with Randel's. They had to convert the 1929 datum into the more recent one, the North American Vertical Datum of 1988, so Randel's data could be directly compared with contemporary elevation data. They trimmed the edges of the DEM so its shoreline—about 10 percent wider than in Randel's era because of landfill—overlapped the nineteenth-century shoreline. Many of the limitations affecting the earlier study persisted, but the direct comparison between Randel's era and today's is fascinating.

Rose-Redwood and Li found that the topography has changed. Differences of as much as 115 feet can be seen in some spots, and in general the West Side is less elevated, the East Side less depressed. The West was smoothed, the East filled in. At the same time, however, the more things changed, the more they stayed the same: "The average change of elevation was remarkably low, the majority of the alterations were only on the order of a few meters or less, and the historical profiles of Manhattan's avenues are very similar to the contemporary transects of those same thoroughfares." The data could go both ways. If readers wanted to argue that Manhattan had been flattened, the evidence was there; if readers wanted to argue that Manhattan by and large has the same swells, the evidence supported that conclusion as well.[27]

For Rose-Redwood the duality of the data demonstrated the power of a frame and interpretive bias. Critical geography, the field Rose-Redwood works in, infuses the study of geography with considerations of power relations and with theoretical frames, bringing Marxist, postcolonial, feminist, and other theories to the analysis of landscapes and people. At root, critical geographers seek to recognize and name the frames imposed by an era, a regime, a society, a scholar, and both to analyze and to lift that particular frame or set of frames. The field of critical geography examines, in part, feedback between ideas and their expression—their physical, material expression—in and on a landscape and the resulting transformations. "Material practice and economic interests influence how we think about things, but how we think about things affects how we organize the spaces that we live in. And then those spaces affect our material interests. So it is this kind of feedback loop," Rose-Redwood summarizes.

Contemporary Manhattanites can discern the primordial landscape rolling beneath their feet; they can be attentive to what endures. Manhattanites can regret the leveling and topographical transformation of the island of many hills. Or Manhattanites can do both at the same time—simultaneously appreciate what has changed, what has not—and consider how they might shape the island's future.

The first topographical reconstruction paper led to Rose-Redwood's talk at the Landmarks Preservation Commission, which led to his encounter with Lemuel Morrison, the New York City surveyor, which led to forays for surveys, which led to the discovery of Randel's bolt. The paper also led to a query from the History Channel, which hoped Rose-Redwood would be interested in portraying Randel in a historical recreation. The two young men were the same age; both were obsessed with the grid. Rose-Redwood declined. But he and Morrison did agree to participate in the television program—*Super City: New York* aired in September 2008—and to reveal the bolt's location, if the History Channel could film without revealing it to their viewers. Rose-Redwood and Morrison were still worried about bolt vandalism.

One afternoon in early 2008, Rose-Redwood met the director on Fifth Avenue near 59th Street and the men made their way into the park. Morrison was running a little late and the film crew had already gone in. At first Rose-Redwood and his companions couldn't find the crew. Finally the two groups located each other. Then Rose-Redwood couldn't find the bolt.

"They are like, 'Where is it?'" Rose-Redwood remembers.

"And the funniest thing was that I couldn't find it! 'I know it is around here somewhere,' I told them. And I call Lem up and he is trying to describe to me some building, some structure, some field. And he was describing it, saying, 'This is where you are, and so turn right.' I took the History Channel people five or ten minutes in the wrong direction." Rose-Redwood is laughing hard as he recounts the story. "Then I called Lem and I said, 'I don't see you.' 'Go back to the building and I will meet you there,' he says. Then Lem couldn't find us for a half an hour. And Lem couldn't leave where he was, because he had all this survey equipment. He ended up running really quick."

Rose-Redwood had experienced that strange dislocation New Yorkers sometimes feel when they come up out of the subway in an unfamiliar neighborhood and uptown, downtown, east, west, have flipped or shifted. The familiar has become unfamiliar. You know where you are and yet you can't find your way.

Surveying the Forest for the Trees

It is a nice coincidence that Lemuel Morrison's first two initials are J. R., because there is much resonance between the two surveyors. Most surveyors are, by inclination and training, precise, exacting in their pursuit of accuracy, interested in instruments. But most of them are not as fanatical as Morrison is. Occasionally, Morrison admits, his engagement with new technology can be too much for some clients, who might not want as much detail as he is willing to—and often does—provide. And few surveyors are as intimate with New York City's landscapes—geological, historical, current—as Morrison. To spend time with Morrison in the field is to see New York unfinished, emergent, and at times as rough and as wild as it can be.

On an August morning a few years ago, Morrison and a crew of two were in a damp grassy enclosure in the northwesternmost section of the Bronx Zoo, hammering numbered tags into trees. The sun was bright, the sky clear, and a lakelike puddle thrummed with mosquitoes. Black squirrels were shrieking, hurling acorns; the resident Sika deer had been led out and into a small corral to the side so Morrison could label and photograph each tree and record its GPS coordinates. "That one is older than New York City, at least two hundred years old," said Morrison, pointing to a towering oak some 13 feet around. A peacock shrilled somewhere to the east.

It was a second survey for Morrison there. Zoo officials hired his company, Mercator Land Surveying, to map the area and to locate trees, particularly those old or rare, that would be important to preserve if any building was done. An arborist identified about eighty different species. But the arborist's identifications and the surveying team's GPS recordings for the various trees didn't align. So Morrison decided to return to get more data and to correct the discrepancies. It was a good-faith effort, he said; he was not being paid for the extra work but wanted to do the job right. This section of the Bronx Zoo's land was unlandscaped, and further north along the dirt road from the Sika enclosure the vegetation was thick and tangled. Bushes, vines, weeds, and brambles stymied the team's attempts to get the GPS unit up close to the trunk of each tree; the canopy wasn't making the satellite readings any easier either—the more cover, the fewer satellite radio broadcasts can get through. The mosquitoes hovered close to the trunks, and the shade whined with them. "At Hunts Point I got seventeen mosquito bites and two tick bites," said Shomari Johnson, one of Morrison's crew, as he swatted. A monarch butterfly drifted by in the sunlight.

Trees, says Morrison, are the bane of his existence. Whereas in Randel's day trees obscured the distance and needed to be cut by ax men so a straight line could be seen, trees today obscure satellite signals, making it impossible to record exact location. Landowners are just as upset about trees being cut as they were in Randel's day, but not because of property claims and shifting borders. Morrison had marked the zoo's trees with tape on his first visit so the arborist could easily see them. Officials seeing pink tape fluttering from tree trunks got the wrong impression. "They thought we were going to cut down all the trees," he explains, laughing. "People got all bent out of shape." While laying down its two hundred growth rings, adding several millimeters each year to its girth, the now towering oak traveled through different cultures and identities. It and its neighbors used to be farm-thwarting or commercial objects. Now they are emblems of valuable intact nature, wilderness, and ecosystem health.

Part of what Morrison loves about surveying is being outside, being in places other people don't get to visit or see, or see in the way he gets to—like that remote part of zoo land, which was readied for new structures after he resurveyed the trees. His work in and around New York City has taken him to Hunts Point, the Brooklyn Bridge, Governors Island, a mucky marsh in New Jersey, a marine transfer station where garbage is put on oceangoing barges, the Hutchinson River Parkway, subway stations, and countless construction sites. Morrison comes from a family of surveyors—and he says there is a slim chance that one of his distant grandfathers, William Crawford, taught George Washington to survey. He has many old instruments that his great-grandfather and grandfather used. "Randel would recognize them," he says. "But my great-grandfather and Randel would not recognize what I use today. Someone from the sixties would not recognize what I use today." One recent October day, Morrison took one of his maternal grandfather's elegant brass theodolites and some heavy wooden tripods into Central Park, set the theodolite above Randel's bolt, and peered through the telescope. He set a backsight on a building on 59th Street and a foresight on a tree between Sheep Meadow and the Mall. With his antique instrument, his small wire-rimmed glasses, and his ruminations about transits and calculating angles of triangles using sine, cosine, and versed sines, Morrison seemed a nineteenth-century apparition. One October day more than two centuries earlier, Randel too stood atop the rock, likely positioning a wooden tripod and theodolite on its rough and striated surface. He marked the spot for a bolt and perhaps readied himself to calculate triangle lengths.

UNTIL THE 1970S, surveying had not changed much since Randel's time. Surveyors still stretched measuring lines—steel tapes instead of chains, but the principle was the same. They used transits and theodolites to observe angles so they could triangulate distances, both horizontally and vertically. They used levels. The development of pocket calculators in the early 1970s began to help with calculations; surveyors didn't have to do the math in their heads, on paper, or with card-reading computers. Then the instruments became electronic, using light in different ways—including laser light—to measure distances. A laser device emits a pulse of light, the color of which depends on the elements used to create the light. The light hits an object and bounces back to the device, which records the time of its return. By calculating how long it took the pulse of light to come back, the device can determine distance. Once the distance information is recorded, computers build three-dimensional images with the data, allowing surveyors to rotate, zoom in on, zoom out of the images. One of the machines that provides these kinds of graphics is called a 3D scanner; not all surveyors use such equipment, because it is expensive to buy, costly to repair, and quickly obsolete. But Morrison loves his 3D scanner.

He used one in the basement of a high-rise going up on Manhattan's East Side not too long ago. Morrison had been hired to survey the basement so the property could be correctly apportioned into residential space and commercial space. Wearing his white hard hat, which seems to permit one unfettered, unchallenged access to active construction sites, he made a mark, a dot with four cardinal lines emanating, on the cement floor with a blue marker. He set his 25-pound Leica Cyrax Laser Scanner on a sturdy tripod directly over the blue dot. Within a minute or so, the dusty walls of the dark basement were slowly, regularly raked, top to bottom, by a line of green laser light that flashed every two-tenths of a degree in a 45-degree swath around the scanner.

"Do I need special glasses?" a workman asked.

Morrison told him not to worry but not to look directly into the scanner. Morrison calls the scanner Apollo. He has given all his equipment Greek names; his two GPS units are Ares and Sirius, his ranger Skylos, his server Sybil. For about twenty minutes the laser captured the dimensions of that section of the room. Morrison then moved the apparatus to another section; he repeated the process ten times, ensuring that he got every corner and section of the sprawling basement from different angles. His crew deftly avoided the laser; they have a honed sense of where they can walk to stay out of the beam and how quickly they need to walk to stay ahead of it or behind it.

While the scanner was recording, Morrison pulled up the images on his laptop. He could see places where the details were not as fine as he wanted or where there were gaps in the data, and then he set the scanner again. He overlaid the images from the different scanning stations using landmarks, or "targets," which allowed him to line them up. For instance, a red knob on a pipe appeared on the far left of one image; that same knob appeared on the far right of the next image, and so the two images could be joined with the knob in the center, and so forth and so on until a panorama of the entire basement was stitched together using landmarks like knobs, light bulb sockets, and a white ball that Morrison set on the ground in a stretch barren of landmarks.

"Shooting a movie?" asked another workman as he passed by.

"That's right. Better smile!" Morrison responded.

In the corner of the room, a worker screwed bulbs into sockets.

"Two more to go," said his assistant.

"Two hundred and fifty-nine more to go," his foreman countered.

As the scanning progressed, Morrison found a problem. The images of the white sphere from two sets of scans didn't line up. "Something got screwed up," he said. "It is almost as if the sphere got kicked. It got kicked." On the screen of his laptop, the two spheres looked out of register, one slightly above and to the side of the other. Morrison decided he could get the software to align the spheres. He could also see a ghost: the faint image of a construction worker leaning against a column in one of the hallways.

Morrison's Leica scanner cost about $130,000, and he says he is one of few surveyors using such high-end equipment for most jobs—an assessment confirmed by Curtis Sumner, executive director of the American Congress on Surveying and Mapping. In fact, Morrison had to teach himself how to use it because none of his colleagues knew how. Without the scanner raking the room one slice at a time, he and his crew would have used handheld scanners and the job would have taken days instead of a few hours. And Morrison notes that the scanner saves more than mere time. He has used it along highways and in subway stations, keeping himself and his crew away from trucks and cars and off the tracks and the third rail. The dangers posed by surveying have changed with the eras. American mosquitoes no longer typically carry malaria or yellow fever. The seasons no longer bring smallpox, cholera, ague, and the usual autumnal fevers. "You read the old surveyors' accounts and it is like 'Thirty-five degrees, sunny, cut down forty-seven trees, two guys killed by Indians today, butchered two mules and ate one,'" Morrison says. "We don't have that. But we've still got the weather."

That Truth Measurement

Aboveground, laser scanning or more basic distance- and angle-observing instruments such as a modern transit or theodolite work in tandem with the other technology that revolutionized surveying and every other location-determining activity: the Global Positioning System. For millennia people looked beyond the bounds of the earth to navigate and to determine position. Today a constellation of man-made satellites orbiting the earth has replaced the stars and planets as guides. The U.S. military launched the first of this network of satellites in 1978 as a means to aid navigation for defense—for soldiers and vehicles, for nuclear warheads and other missiles. In the 1980s, GPS became available for civilian use, with some security provisions built in. The system is based on precise knowledge of orbits and on radio signals broadcast from twenty-four satellites circling 12,500 or so miles above the earth's surface (three additional satellites serve as backup). The GPS receivers people use in everyday life—in their cars, when hiking—receive information on specific frequencies, called L1 and L2, from at least four satellites and have an accuracy of about 10 to 20 meters. (GPS measurements are usually metric.) Higher-end GPS receivers, such as those used by surveyors, can pick up the signals of eleven to twelve satellites, and when those data are combined with data from GPS ground stations they can give location to within several centimeters. Neither the general public nor surveyors can interpret additional military-only signals. A new generation of satellites will broadcast on another frequency, L5, "and that frequency will greatly improve the capability of GPS," says David Doyle of the National Geodetic Survey. The U.S. system is not the only constellation: Russia has GLONASS, the European Union is setting up Galileo, and China is launching its own.

All GPS satellites circle the earth twice in twenty-four hours, and their orbits are designed to ensure that a minimum of four satellites can be accessed from any place on the planet's surface at any time. Each satellite broadcasts its own signal, which announces its identity, what orbit it is on, and the precise time according to an onboard atomic clock. Atomic clocks give a precision that a regular quartz clock cannot attain; they are based on signals emitted by electrons moving regularly between different energy states. An atomic clock varies, at most, one second in 20 million years.

Just as time and an accurate clock were the keys to determining longitude in the eighteenth century, time and an accurate clock remain the keys to determining location today. Trimmed of a few details, the process works like this: Both the GPS receiver and the satellite emit a regular synchronized code, but travel time from the satellite makes the arriving code slightly off-beat with the receiver's code. The difference in the beat—say, seven-hundredths of a second—permits the receiver to calculate how far the signal traveled. (The formula is velocity × time = distance. Velocity is about 300,000 kilometers per second, or the speed of light, and in this example time is seven-hundredths of a second. Therefore, the first satellite is 21,000 kilometers away.) The arriving signal also declares the correct time at its moment of departure from the satellite; the receiver can correct for the delay once it knows the satellite's distance and can calibrate its quartz clock to the first satellite's time. A second satellite signal arrives, and its time and code are, of course, slightly different, because they too have been delayed as they voyaged through the atmosphere; that delay of hundredths of a second permits the GPS receiver to calculate how far away the second satellite is. And so on. When the receiver resolves its distance from four satellites it can use trilateralization—essentially triangulation in three dimensions—to calculate position. Each satellite can be thought of as sitting at the center of a sphere. The distance from the satellite to the GPS receiver is the radius of each sphere. The point where all four spheres intersect is the point where the GPS unit sits.[28]

"The system started going up in '78 and '79, and by the mid-eighties, we could measure across the United States with centimeter accuracy. So that completely revolutionized surveying measurement," Morrison explains. The level of precision is so great that surveyors and others using GPS must note the date of their measurements because of continental drift. "We are moving away from Europe and we can measure it. We move about a centimeter every year. So you have to say, 'This is the position as of this date.' The whole globe is moving and subsiding and rising, and we can measure it now. But," Morrison cautions, "when you can measure better and better, that truth measurement is elusive. You say something is two meters long. Well, it is not exactly two meters long. It might be 2.00001 or 2.00002 or 2.00003."

To correct for some of the known errors that arise, the satellite network has a ground component. Errors are introduced because of special relativity (clocks on the fast satellites run more slowly than clocks on earth), solar flares, and atmospheric effects—some of them the same undulations that delayed Randel. Water vapor, heat, and dust still play the same tricks. But the ground network can discern how the satellite signals are affected. Called Continuously Operating Reference Stations, or CORS, and managed by the National Geodetic Survey, these GPS receivers receive positioning information 24/7, which means that their locations are extremely accurate. Surveyors standing in the field can watch their GPS coordinates constantly shifting slightly. "I am not moving, but it says I am," Morrison explains. So he can dial into a nearby CORS station, which has averaged its position for, typically, years, making it spatially resolved. It can tell Morrison what mistakes the GPS is making, and then he can correct for those errors. "That is what is called survey-grade position," he says. "Accuracy within five centimeters."

Morrison regularly relies on several CORS in and around New York City. The New Jersey Institute of Technology has one on its roof. The North River facility just off Manhattan at West 137th Street has one (problematic because it bobs a few centimeters up and down on the tide). Queens has one. There are roughly 1,800 of them around the country, some run by the National Geodetic Survey and some by state agencies, companies, or academic institutions. Many nonfederal CORS share their data with the National

Geodetic Survey, which maintains the information and "gives it the seal of approval," says David Doyle.

The CORS are part of the National Spatial Reference System, which has its origins in Randel's time. "It is the most important part of our national infrastructure," says Doyle. "Think of it as the foundation of a huge building. Once it is laid down, you can forget about it. You can change the building any time you want. Put on new walls, a new roof. You can do all that without changing the foundation." The system originated with the U.S. Coast Survey, which Thomas Jefferson established in 1807 and is the U.S. government's oldest scientific organization. Ferdinand Hassler—the Swiss surveyor who did not get the Manhattan job eventually given Randel—directed the Coast Survey off and on for many years, starting in 1816. While Randel was up and down and back and forth across the island, striving to bring mathematical precision to the grid, Hassler was up and down the Hudson River and back and forth between New Jersey, Staten Island, Long Island, and Manhattan, establishing a series of markers at the points of triangles that were the first national geodetic markers—the first points giving rise to a network that culminated in the establishment of the national datums. One of Hassler's markers survives at the Sandy Hook lighthouse in New Jersey.

Hassler's measurements were the basis for America's understanding of the size and shape of the continent for nearly half a century. Between the 1870s and the late 1890s, surveyors measuring the West Coast realized that the shape of the East Coast was not prescriptive for the continent, and a new model for the size and shape of America was derived. It came to serve as the model for Mexico and Canada as well. Alaska and Hawaii had their own datums. Europe and Britain had theirs. Japan too. And India. Each swath of the earth had a view of what the earth looked like "that best fit that piece of geography," explains Doyle. Each was based on regional measurements that gave ever-so-slight differences in estimated values for the distance between the center of the earth and the poles (called the polar radius), the distance from the center of the earth to the equator (equatorial radius), and the flattening of the earth at the poles. "There was a grand idea that there would be this unifying concept," Doyle says, "but historically there have been in excess of fifty and maybe between sixty and sixty-five different mathematical models for the size and shape around the world." Between 1830 and 1984 some of the models included Airy, Everest, Bessel, Clarke, Helmert, Hayford, Krassovsky, Fischer, WGS 66, IUGG, WGS 72, International, GRS 80, and WGS 84. Sometimes the model had nothing to do with the landscape or the region but with politics and secrecy. "If you look at Europe following World War II, the NATO countries adopted the European Datum of 1950. All the eastern bloc countries adopted Krassovsky 1940." Doyle laughs. "Are you telling me that if I make one step from Austria into Czechoslovakia, the size and the shape of the earth change?"

Today the ellipsoid, as the figure of the earth is called, is still being refined. When scientists placed the datum in the earth's center, they could do so with an error of plus or minus 2 meters. "Which is not bad," says Doyle. "It is a big planet." In the thirty years since, the earth's center has been located to within 3 centimeters of error. Doyle says eventually the datum will have to be shifted again and everyone's maps and charts and positions will have to be converted to the new datum. "It would change everyone's position by one to two meters. You can know exactly where you are and still be lost." And that change would just entail the horizontal datum. The vertical datum will have to change too. In ten or twelve years, measurements of gravitational fields taken on the ground, in aircraft, and by satellites will yield a new national standard.

"People are becoming very accustomed to knowing where they are. Everything is about place. Everything revolves around where you are, where you want to be, where you have been," Doyle says. "It used to be that when my office phone rang it was a surveyor, or my wife. No one else cared about this information. Now it is John Deere and farmers involved in precision agriculture. It is people in architecture and construction. It is folks doing marsh rehabilitation who want to know the height of the mud to a millimeter. We are just seeing so many new users; we don't know their jargon and we don't know their jokes. This is a real challenge for us: to build and maintain this reference system that maintains the legacy."

Hassler and Randel are among the grandfathers of that legacy. They loved precision and accuracy. Both were confident that surveys done with mathematical rigor were the core of a society's success; they measured the present to ensure the future. Surveying monuments from their era are among the more than one million vertical and horizontal markers, all referenced to one another, that make up the National Spatial Reference System. Markers have been embedded in pavements, in walls, in boulders; one, set in 1965, sits in Big Rock in Riverside Park, where children have happened upon it and wondered about the shiny metal disk. Without the National Spacial Reference System and GPS, there would be no cell phones, no electronic banking, no air traffic control, no GIS models, no synchronized computer networks, no Google maps, no GPS receiver in your car coolly, dispassionately telling you where you are and where you should go.

IV

In Which Randel Keeps Seeking the Most Eligible Routes

"This day sent out through the post office about 160 copies, being all that remained on hand, of the first Edition of The Exhibit of the oppression of John Randel. Three of them were returned by the post office, whether the parties were not to be found, or refused them, I cannot tell." So wrote Mathew Carey, the prominent Philadelphia publisher, in November 1825. Carey defended Randel in one of the most publicized, longest-running, most financially punitive lawsuits of the era, and as he wrote this entry, the legal struggle was just beginning. Carey printed and distributed several pamphlets—most famously, *Exhibit of the shocking oppression and injustice suffered for sixteenth months by John Randel, Jun. Esq.*—in unwavering public support of Randel, despite, as he noted in the same entry, potential apathy on the part of his audience. Undeterred, Carey's highly visible outrage was costly: "I have spent my money to a great extent (about $150), made many deadly enemies, exposed myself to obloquy." But, he added, "I do not regret it nonetheless. It is one of the best acts of my life."[1]

This lawsuit, between Randel and the Chesapeake & Delaware

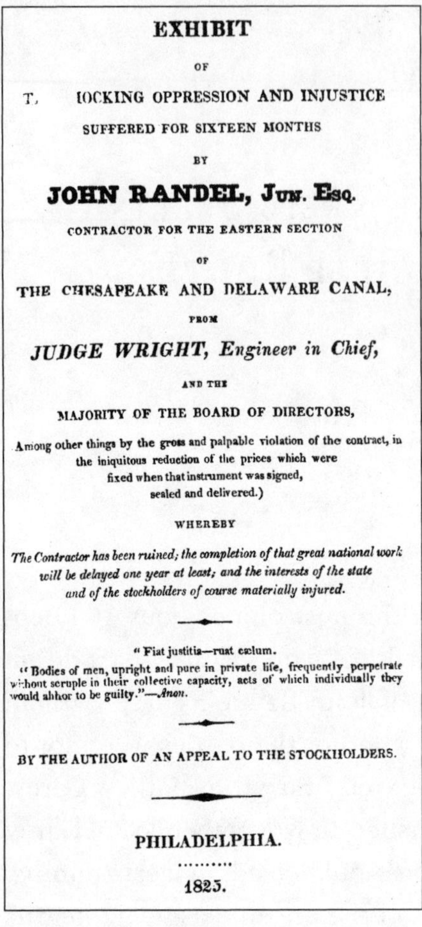

EXHIBIT

OF

T, IOCKING OPPRESSION AND INJUSTICE

SUFFERED FOR SIXTEEN MONTHS

BY

JOHN RANDEL, Jun. Esq.

CONTRACTOR FOR THE EASTERN SECTION

OF

THE CHESAPEAKE AND DELAWARE CANAL,

FROM

JUDGE WRIGHT, Engineer in Chief,

AND THE

MAJORITY OF THE BOARD OF DIRECTORS,

Among other things by the gross and palpable violation of the contract, in the iniquitous reduction of the prices which were fixed when that instrument was signed, sealed and delivered.)

WHEREBY

The Contractor has been ruined; the completion of that great national work will be delayed one year at least; and the interests of the state and of the stockholders of course materially injured.

"Fiat justitia—ruat cœlum.

"Bodies of men, upright and pure in private life, frequently perpetrate without scruple in their collective capacity, acts of which individually they would abhor to be guilty."—*Anon.*

BY THE AUTHOR OF AN APPEAL TO THE STOCKHOLDERS.

PHILADELPHIA.

....

1825.

Exhibit of the Shocking Oppression by Mathew Carey.

Canal Company, as well as several related lawsuits, unfolded over more than a decade. The cases laid bare the intense politics, personal rivalries, and finances that underlay the building of America. And they flayed Randel, shattering his finances, revealing disturbing facets of his personality. It was with the Chesapeake & Delaware Canal lawsuit and the many that followed that he began to earn his reputation as an erratic, litigious man. Increasingly, Randel's vision for the future—his opinions about the routes and engineering of canals, the routes and mechanics of railroads—brought him into conflict with some of his contemporaries, most notably, in the case of the Chesapeake & Delaware lawsuit, Benjamin Wright, a judge and surveyor turned engineer from Rome, New York. Despite his legal and economic setbacks and cautions from employers, Randel became uncompromising in his middle age. He held to his visions, his calculations, and his convictions with an obsessive fervor.

Randel's trouble with the Chesapeake & Delaware Canal began on America's ur-canal, the Erie. His unrelenting, unabashed quest for precision led him to miscalculate, politically and professionally.

THE ERRONEOUS DREAD OF BOLD MEASURES

Just as the young United States established property lines so land could be claimed and "improved," so it needed to establish connections between those improvements, channels of transportation and communication, themselves "improvements." The term was the watchword of the late eighteenth and nineteenth centuries. Improvement was *the* American enterprise, as Basil Hall, a Scottish captain who traveled through New York State in the late 1820s, aptly described:

> It may be proper to remark, that about this period I began to learn that in American the word improvement, which, in England, means making things better, signifies in that country, an augmentation in the number of houses and people, and above all, in the amount of the acres of cleared land. It is laid down by the Americans as an admitted maxim, to doubt the solidity of which never enters any man's head for an instant, that a rapid increase of population is, to all intents and purposes, tantamount to an increase of national greatness and power, as well as an increase of individual happiness and prosperity. Consequently, say they, such increase ought to be forwarded by every possible means, as the greatest blessing to the country. I do not assert that Americans are entirely wrong in this matter; far from it; increase of population is sometimes a symptom of prosperity.

But, he added, "Much of the wealth, and power, and happiness of nations have their origin, and still more their permanent support, in circumstance of which little or no account is taken in America."[2]

Over several decades the meaning of "improvement" evolved. Initially, as Hall noted, it was a blanket term encompassing everything from clearing land and expanding population to building structures and institutions. When, at the turn of the century, DeWitt had instructed his surveyors to note and assess every improvement, he typically meant deforested or farmed land and built structures. But by the mid-nineteenth century "improvement" came to mean, more narrowly, public transportation.

The growing focus on transportation had its official roots in 1807, the year the commissioners started work on the New York City grid, when Congress asked the secretary of the treasury to study and report back on the possibilities, costs, and challenges of national infrastructure. Albert Gallatin issued his *Report of the Secretary of the Treasury; on the Subject of Public Roads and Canals* in 1808. Gallatin started by noting that "The general utility of artificial roads and canals, is at this time so universally admitted, as hardly to require any additional proofs." He then furnished additional proofs, emphasizing that "no other single operation, within the power of government, can more effectually tend to strengthen and perpetuate that union, which secures external independence, domestic peace, and internal liberty." Without a network by which people, goods, and information could move, politicians and business leaders realized that there might be no further "improvements" and that the newly assembled states would not become tightly stitched into an ensemble.[3]

Gallatin's report reviewed each region, describing the setting, geography, and scope of projects under way or required. He argued that state or local projects needed to be linked in one great inland system. Attached to the report were two supporting statements: one from Robert Fulton, the inventor and businessman, who was then testing one of the earliest American steamships on the Hudson River, and one from Benjamin H. Latrobe, a well-known architect and engineer, who had devised in 1798 a means of bringing fresh water to Philadelphia through steam-driven pumps. Fulton, a published canal booster, described how great civilizations advanced from roads to canals. Although the United States needed

Internal Improvements. Detail from the 1828 "Maps and Profiles of the Canal Line on the North Branch of the Susquehannah River from Nanticoke Falls to Northumberland." Record group 17, records of the Board of Canal Commissioners. Courtesy of the Pennsylvania State Archives.

both, particular attention must be paid to inland waterways, he contended, with a tinge of utopianism: "There is another great advantage to individuals and the nation arising from canals, which roads can never give. It is that when a canal runs through a long line of mountainous country, such as the greater part of the interior of America, all the ground below for half a mile or more may be watered and converted into meadow and other profitable culture. How much these conveniences of irrigation will add to the produce of agriculture and the beauties of nature, I leave to experienced farmers and agricultural societies to calculate."[4]

Latrobe, for his part, reviewed practical details attending the specific projects and raised a critical concern: the shortfall of engineers. "The difficulty of carrying canals parallel to our great rivers, the scarcity of engineers possessing knowledge and integrity, the want of capital, and above all the erroneous dread of bold measures, and the fear of uselessly expending money in works hitherto unknown among us, has deterred those interested in improving our navigation, from deserting the beds of our rivers, while it was practicable to keep them." The word "integrity" appeared often in recommendations for engineers, in discussions of reputation and skill; Randel's letters of support, for example, often described his integrity. A valuable quality in any era, integrity was notably coveted at a time when so few people knew what they were doing or had any means of vetting a man or his services. Financiers, civic bodies, and the public were all vulnerable to huge losses on engineering projects, short-term and long-term. Few national proof-of-concept canals or, later, railroads existed; most models lay across the ocean in markedly different landscapes. ("As many Americans were willing to point out, everything was bigger, better, and deeper in America; the mountains were higher, the frosts more intense, the snows deeper, the spring freshets more violent," notes Julius Rubin in *Canal or Railroad?*) Techniques and true understanding of terrain and physical hurdles often arose on the job. As a result, estimates of cost could be, and typically were, off by hundreds of thousands of dollars. Integrity was thus a vital quality for civil engineers or surveyors. Without integrity—as judged by their church, their mentors, their colleagues, their former employers—they would not survive professionally.[5]

One of the canals Latrobe reviewed in his remarks and Gallatin cited as integral to nationhood and the internal network was to be in New Jersey, running between the Delaware and Raritan Rivers, linking Philadelphia and New York City. Such a canal would obviate a longer trip via the Atlantic Ocean, the Chesapeake Bay, and the Delaware River. Rather, boats could go directly from the base of New York Harbor into the Raritan River and then, via canal, to the Delaware River, just northeast of Philadelphia, cutting about 160 miles and several days off the journey. Randel fastened his hopes on this canal as his chance to raise himself from surveyor to something grander. He hoped to reach the lofty perch of civil engineer, specifically one instrumental to the great national work of infrastructure- and nation-building. An intelligent gamble, but a gamble Randel lost.

Randel's involvement with the Delaware & Raritan Canal arose from his New York City work. As he was setting monuments and bolts on the island, John Rutherford, one of the commissioners of the 1811 street plan and a former New Jersey senator, hired him to survey a route for the proposed canal. Rutherford and others had long advocated such a canal, an idea that William Penn, the founder of Pennsylvania, had floated as early as the 1690s. In 1816 the New Jersey legislature gave Rutherford and two other canal commissioners authority to survey and map a potential route. They in turn hired Randel. He was to "pursue a level line as far as was practicable from Longbridge farm to the Delaware, and to the Raritan, in the shortest direction that the ground would admit, which line should be run with the greatest accuracy, and be esteemed the base line of the work." Between October and December, Randel surveyed, set monuments, and recorded elevations. He was also required to do something no record attests to his having done before: he had to estimate water flow. No description of his method survives, but he was assisted by a millwright, for whom correctly estimating water flow was livelihood. It provided good training for Randel's future work. Ensuring that canals had an adequate supply of water and could sustain the depths required for barges was crucial. Randel and the millwright concluded that the canal and locks would need less than one-eighteenth of the water running through the myriad streams along the route, leaving more than enough for the area's many mills.[6]

Soon after Randel finished mucking through marshes and estimating bank slope and water flow in New Jersey, he received an enticing offer from New York State. The commissioners for the proposed Erie Canal, many of them Randel's acquaintances, offered him a position as an engineer—an engineer on an extremely ambitious, technically unprecedented, and politically complex venture.

Visionaries and businessmen had long contemplated a canal running 360 or so miles from Lake Erie to the Hudson River. A trade and travel route wove along the same path well before Europeans arrived on the continent. Arguments in favor of such an extensive canal were varied, but among the most compelling was proponents' observation that traffic and goods from the west would increasingly run through the Great Lakes and then across Canada to the Atlantic Ocean if New York didn't intercede. "It is evident that the canal will, if properly effected, turn to the United States the commerce of the upper lakes," wrote the first group of Erie Canal commissioners in 1810. Despite its length, such a canal seemed feasible. The Mohawk River Valley stretches east-west, a low-slung band wending between the Catskill Mountains to the south and the Adirondack Mountains to the north. The valley was roughly level; the greatest elevation, about 570 feet, occurred near Lake Erie, and the rest of the route was, so to speak, downhill. Gouverneur Morris, Simeon DeWitt, DeWitt Clinton, James Geddes, businessman Elkanah Watson, and politician Stephen Van Rensselaer were among those who first advocated for the canal. And in early 1811, it was Morris who—just as he signed and placed his wax seal on Randel's map of New York City—was tapped to present a convincing case for funding to Congress. Federal financing was not forthcoming, however, and so the state shouldered the cost, amid many objections. "Ironically, the grand improvement that would fix forever New York's national commercial hegemony prevailed against the opposition of every single delegate from New York City," writes John L. Larson in *Internal Improvement*.[7]

A great deal has been written about how the Erie secured that commercial hegemony and about the environmental and economic changes the canal entrained. Land along the route followed a transformation the Scottish visitor Hall vividly portrayed: "We reached the village of Syracuse, through the very center of which the Erie Canal passes. During the drive we had opportunities of seeing the land in various stages of its progress, from the dense, black, tangled, native forest—up to the highest stages of cultivation, with wheat and barley waving over it: or from the melancholy and very hopeless-looking state of things, when the trees are laid prostrate upon the earth, one upon top of another, and a miserable log-hut is the only symptom of man's residence,—to such gay and thriving places as Syracuse; with fine broad streets, large and commodious houses, gay shops, and stagecoaches, waggons, and gigs flying past all in a bustle." The waters of the artificial river came to fertilize a cultivated landscape along 364 miles as they flowed and stepped up and down eighty-three locks.[8]

On that waterway, goods moved quickly. Markets opened up, and tolls along the canal added up. One historian has calculated that in 1800 four horses could pull one ton 12 miles on a typical road over the course of one day—a road often deeply grooved with ruts, pockmarked with deep holes; a road of dirt, gravel, rough-cut stone, or bumpy with tree trunks placed one after the other. ("Throughout the nineteenth century, European travelers rated American natural roads as the worst in the world and American turnpikes little better," describes one scholar.) On the canal, in contrast, a team of horses could pull thirty tons 24 miles in one day. By one account, costs for a ton of freight moving from Buffalo to New York City fell from $100 to $10 and travel time fell from twenty-six to six days. Within nine years of the canal's completion in 1825, tolls recouped the $12 million or so spent on planning and construction. Albany and other upstate towns boomed, and New York City moved into position as the major U.S. port. Goods flowed readily between the east and the frontier as it pushed west.[9]

Much has also been written about how the many engineers of the Erie Canal became the leading engineers of their day, although none had trained as such. Benjamin Wright—Randel's future antagonist—had a typical story. He was a judge who had done some surveying, at times quite shoddily, and who then

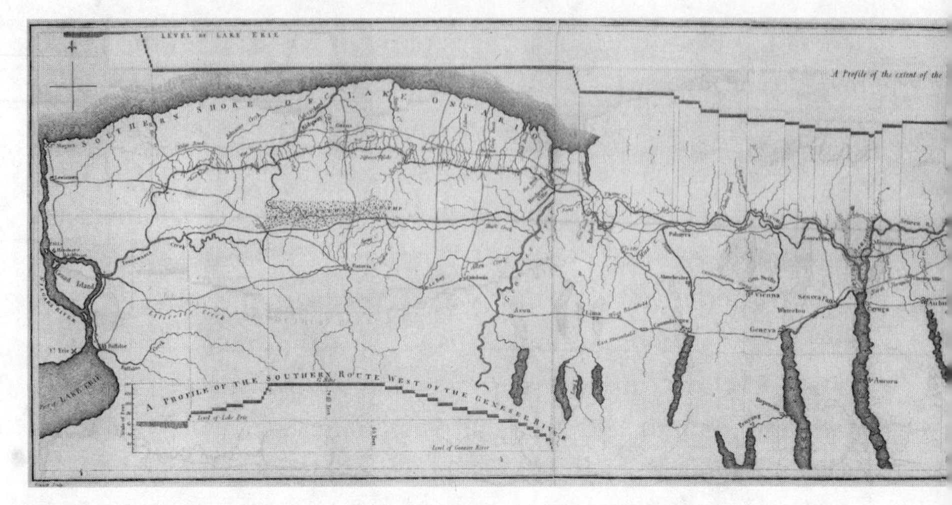

1817 map of the proposed Erie Canal. Courtesy of the Erie Canal Museum, Syracuse, New York.

taught himself engineering on the job. He and his colleagues studied European texts, and one of the Erie engineers, Canvass White, made an observational visit to England, walking some 2,000 miles to examine the British system. Because of their rite of passage on the longest, most successful canal of that time, these men were subsequently sought out by other canal commissions—in Connecticut, Rhode Island, Delaware, Maryland, Virginia, Pennsylvania, New Jersey, and Illinois—for their expertise and for the techniques they devised and tested. They invented devices to pull down massive trees and to wrench stumps and roots from the ground; they concocted a local cement. They became coveted experts during the canal boom.

In 1817, when state funding for the Erie Canal came through and digging began, the long-term professional benefits of Erie employment were perhaps not yet so clear, whereas the future of the Delaware & Raritan Canal seemed assured. Randel turned down the New York State commissioners' offer to be engineer for the eastern third of the Erie, which was to run from Utica to the Hudson River. Their choice of him for this section of the canal revealed that they "had equal confidence in my capacity

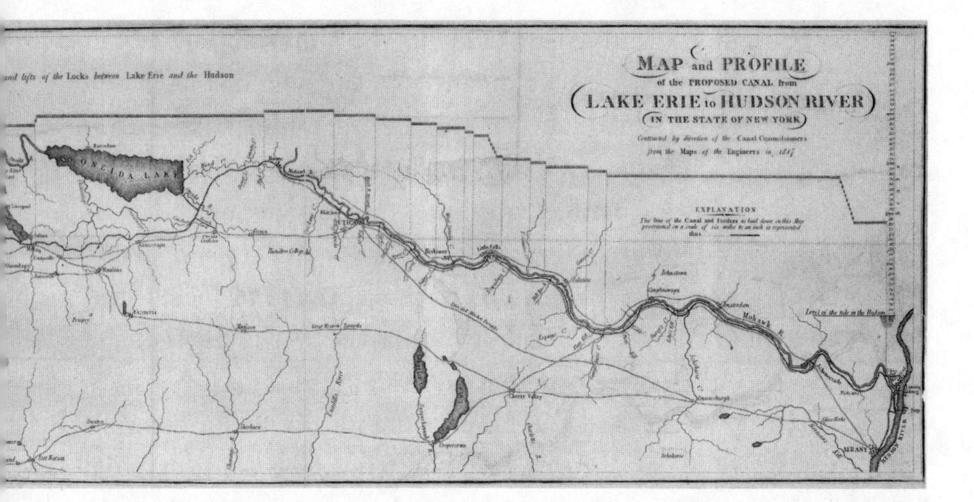

with that of Mr. Wright and Geddes," he wrote in a draft letter to Simeon DeWitt. Wright had been appointed engineer in chief of the central third and later oversaw the eastern third. James Geddes, who had surveyed some of the upstate land Randel had purchased, was in charge of the westernmost portion of the canal, which ran from the Lake Erie terminus at Buffalo to the Seneca River. DeWitt must have concurred with Randel's assertion of confidence, for he described the eastern section as "the most difficult part of the canal." There, for 30 miles between Schenectady and the Hudson River, the Mohawk River Valley became particularly steep and circuitous, and elevation fell sharply. DeWitt later wrote that Randel turned down the appointment for "reasons of a private nature." Perhaps he declined the Erie because he was in charge on the Delaware & Raritan; as chief surveyor he may have anticipated that appointment flowing into an appointment as chief engineer. Whatever the reason, Randel made the wrong choice.[10]

Despite Gallatin's report, opposition to canals cut deep in many places, and the "erroneous dread of bold measures" was quite alive in New Jersey; two attempts previous to Rutherford's and his cohorts' had failed. Conflict arose about the water supply, about how the canal should be funded, about who and which state would most benefit. Such opposition and conflict was typical. As Larson explains, "Very quickly—and with no little irony—people found the seeds of paralyzing conflict within the broadly popular and virtuous objectives of internal improvement. They took to supporting or opposing public works according to their private or local interests, and sometimes according to whether they believed that the revolution had empowered the nation, the states, or private individuals to impinge on the conditions of life for the larger community." Because of the ongoing debate, the Delaware & Raritan Canal lay dormant until 1830. The 44-mile canal was finally completed in 1834, nearly twenty years after Randel surveyed the route.[11]

By 1820, Randel knew he had misjudged. He was not among the men emerging as famous, sought-after engineers on the Erie. He fretted and obsessed about the consequences of his decision. But neither Randel's personality nor his personal economy could tolerate obscurity, any more than they could tolerate obloquy. So with some well-placed aid, Randel inelegantly inserted himself in the foreground. He compounded one unfortunate professional decision by making several worse ones.

No longer needed in New York City or New Jersey in early 1821, Randel traveled to Albany with Matilda on a Hudson River steamboat ($12, he noted), sending his horse and wagon separately on a sloop ($5). With his father's help, he colored and varnished his newly engraved New York City maps and settled into tending the garden of his Albany home at 168 North Market Street, hiring workers to plant and to construct a stone wall. (Now Broadway, North Market Street was one of the central and original streets of Albany and ran parallel to the river.) Randel loved his garden, and even from afar he cared for it, writing letters to caretakers to ensure that plants were well protected. One winter, for instance, he issued detailed instructions for the care of his fig, madeira, and walnut trees.[12]

On May 16, Hermanus Bleecker, a powerful Albany lawyer, visited Randel and asked him to survey "without instruments" (that is, cheaply and informally) a route for the easternmost part of the canal, for which "they the citizens" would pay $25 or $30 a day. The following day was rainy, but Randel traveled in a stagecoach to Schenectady, about 15 miles northwest of Albany, along the turnpike he had surveyed when he was eighteen. Construction of the eastern third of the canal was proceeding apace eastward from Rome and would soon reach Schenectady. At that point the commissioners would need to make a final decision about exactly how the canal would run between Schenectady and the Hudson River—a decision with important economic consequences and thus a controversial one. The general view held that from Schenectady the canal would run along the south side of the Mohawk River Valley to a town with spectacular waterfalls, Cohoes Falls—close to where the Mohawk empties into the Hudson—and then down to Albany. A small feeder could link the Erie Canal at Cohoes Falls to the Hudson River just opposite the city of Troy, on the east side of the river. But another idea had been raised as well: that the canal would at some point cut out of the Mohawk River Valley and travel more directly to Albany, which was then 6 miles south of Troy. If the canal diverted to Albany, Troy would not be the effective terminus and would be deprived of revenue from the canal traffic. Citizens in both towns agitated about the route. Randel went over the land alone for several days and then met with Bleecker and three other men, including "Dudley," probably Charles E. Dudley, the mayor of Albany. He showed them "the canal ground." The following day was a Sunday, and Randel did something rare: "Remain home, have a relax."[13]

Then, for a total of six days, he conducted a more detailed rapid-assessment survey. Accompanied by three assistants, Randel moved from peg to peg along a proposed route, starting at canal peg No. 4774, which "I call No. 1 odd." He noted meadows, houses, winding paths, and the depths of creeks, sketching some of them in cross section. He notched an apple tree; he notched a cherry. The minimalist notations in his field books do not convey the rough character of the land he was hiking, although it is sure he felt the physical challenges of the elevations and steep embankments of the Mohawk River Valley. He must have carefully observed the powerful falls at Cohoes too, considering how best to run a canal through this terrain, the roughest—and perhaps the most beautiful—of the entire route.[14]

After his survey, Randel returned to Albany, worked on his calculations and, again, his garden. On June 8 he went back to Schenectady, this time to meet with the canal commissioners and engineers to talk about the Schenectady-Albany route. The following day, a Saturday, he called on the mayor of Albany to submit his account (for a total of $112.50: "only $100 being appropriated by them the Corporation of the City of Albany . . . the remainder to be paid by individuals").[15]

A few weeks later, after finishing this special assignment for Bleecker and the "respectable body of citizens," Randel went public with his survey findings. On June 27, Stephen Van Rensselaer, one of the canal commissioners and a powerful, rich Albany businessman, introduced in the *Albany Daily Advertiser* a report in which Randel presented conclusions based on his rapid survey. In his report, entitled *Description of the Route of the Erie Canal from the Mohawk to Hudson's River Explored by John Randel, Jun. with Calculations of Its Comparative Advantages*, Randel said he had studied the land east of Schenectady to determine the best route. He proposed an alternative to following the Mohawk River Valley all the way to the Hudson River: he suggested a 3.5- to 4-mile shortcut in the form of a tunnel ("through ground apparently consisting of yellow clay," which would provide the requisite brick). It would run just south of the Mohawk Valley to Glen's Creek, about 3 miles west of the Hudson River. He pointed to the Blisworth Tunnel on the Grand Union Canal in Northhamptonshire in England as a model. A tunnel along his proposed route, he argued, would trim 7.5 miles and cut transportation costs by 30 percent, saving merchants $114,750 a year on freight. A positive assessment from Philip Hooker, a famous Albany architect and a member of the Society for the Promotion of Useful Arts, about the feasibility and cost of such a tunnel followed Randel's report.

Randel's efforts in Manhattan had pushed beyond official requirements because of his exactitude and attentiveness to accuracy. And his creation of instruments had revealed his deep interest in mechanics and invention. His tunnel proposition reflected both facets of his personality. As Simeon DeWitt noted in a short introduction to Randel's newspaper report, "Tunnelling is a novelty in our country, but the commissioners well know that, in Europe, it has been adopted in many instances without hesitation, and practiced in form to more than double the distance here required." Randel was not the first to suggest a tunnel for the canal. Such a proposition had been raised in an 1810 canal commissioners' report: "It may become, therefore, in many cases, more advisable to pierce the earth by a tunnel, than to take down the top of a hill." It had been discussed as a possibility on the south side of the Mohawk River Valley too. But by describing it in detail in the newspaper and including an architect's positive review, Randel introduced the concept to many New Yorkers in an authoritative way.[16]

Randel's *Albany Daily Advertiser* report accomplished several ends. The public now associated his name with the Erie Canal, and with the positive opinion of one canal commissioner, Van Rensselaer, and one well-respected leader, DeWitt, both of whom desired the canal to favor Albany over Troy. And he had proposed a solution worthy of respect—an ambitious innovation, one requiring engineering skill and experimentation, one revealing sophisticated knowledge about advanced English engineering and infrastructure.

I Should Be Pleased to See You, or Receive a Line from You

Whether Randel's newspaper report successfully established further credibility is unknown. He certainly aggravated a number of commissioners and engineers and ignited even livelier discussion about the particulars of the eastern route. The commissioners and engineers met in Albany to discuss alternatives on July 13. A few days later, at 4:00 in the afternoon of July 16, Randel went to James Geddes's lodgings to hear "his objections etc. to my plan." Other Erie men were also in the room: William C. Young, William C. Bouck, Myron Holley, and Canvass White. (Young was a surveyor, Bouck a commissioner, Holley oversaw construction and was treasurer, and White was the engineer who had conducted canal reconnaissance in England.) After the meeting, which lasted more than an hour and a half, Randel rushed home to write about it: "I have this moment returned and about 15 minutes ago the above was said and I now commit it to writing that I may be correct." Randel recorded asking Geddes when a decision about the eastern route would be made, making clear his own predilection: "Albany would not be willing to give up till they must." Randel then quoted Geddes's response: "They who will publish letters to the public must bear the consequences, for his part he did not intend to descend to particulars, but for his part he thought it would have been a disgrace to him or any of the engineers to have made such a statement. I turned to Mr. White and asked him if that was his opinion also he replied yes." In other words, both men rebuked Randel for publishing his survey findings and his proposed plan and for thrusting an alternate route into public view. They were of the opinion that Randel should have been more diplomatic.[17]

The canal route east from Schenectady remained unsettled.

For the rest of the summer, Randel surveyed for DeWitt in

the Onondaga Salt Springs, the town of Salina, and elsewhere. His presence in upstate New York encouraged and facilitated his interest in the Erie, to no good end. In mid-August he briefly met Benjamin Wright and showed him his idea for the route. Wright was, per the commissioners' orders, reviewing all options. He apparently said to Randel then that "he had never examined this route," meaning Randel's. A few days after their tête-à-tête, Randel copied notebooks in the possession of Daniel Judson, one of Wright's assistants, and "shewed him the whole route." By copying Wright's records, Randel got a clear sense of Wright's ideas and some of Wright's calculations. By this point a third alternative had been suggested: after Schenectady the canal would not travel along the south side of the Mohawk River through the valley but would cross the river and use the northern side of the valley, which appeared to have more stable ground.[18]

Late September brought Randel's next political misstep—the seed of the Chesapeake & Delaware Canal lawsuit. On September 19, under the heading of "Calculations of Direct Route of Erie Canal from Schenectady to Albany," he entered into one of his field books figures for the amount of soil needing removal and the cost and time it would take, based on the calculations Wright had made for the commissioners for a cubic foot of sand: ". . . then 27 cubic feet, or 1 cubic yard, weighs . . . 1.182 tons. 1 horse will draw in a scow 50 tons 2 miles an hour & return empty 2½ miles an hour average per load 2¼ miles per hour. To manage this load will require 1 horse at a daily cost of .75 cents, 1 Rider at a daily cost of .75 cents, 1 at the helm at a daily cost of .75 cents. daily cost per load = $2.25. This horse will haul 10 hours per day." And on it went. A stream of mathematical consciousness. The calculations prompted Randel to communicate with Wright.[19]

Randel's draft letter of September 28 sounded collegial. He remarked that he had found the cost of removing the earth "will be much less than you or I estimated." He continued, using "we" frequently. He concluded that the route he had proposed would be cheaper and could be completed faster than any other. And, he added, "I am particularly desirous of seeing you before I publish my calculations"—perhaps for professional courtesy, perhaps respect for Wright, or perhaps a better appreciation of the political fallout of publishing without advance warning. Randel apparently did not hear from Wright expediently, and composed another letter to him a few days later, suggesting they meet in Rome, wherever convenient for Wright. "Will you therefore have the goodness to leave or send a line to me at Utica, to be left at Bagg's Tavern," Randel requested. "I should be pleased to see you, or receive a line from you."[20]

Randel continued surveying for DeWitt through the cold winter. He worked with a crew of chain men, ax men, and a flag man. The men wallowed through boggy sand and spongy swamp. At one point Randel sprained his leg and his frequent partner in the field, Charles Brodhead, who had been working on the canal but was now surveying again in the new towns, became severely ill. In late November snow and moisture interfered with the needle of the compass and the crew had to stop for a half a day. Randel spent his thirty-fourth birthday in the field, away from Matilda. He spent Christmas in the field as well.

Randel included recipes in his notebooks, and several notable ones were recorded in this region or during this season. Even in miserable field conditions, he sometimes ate and drank well—or, just as likely, his thoughts turned to eating and drinking well.

Small Beer

(recorded in Liverpool, New York)

2 gallons molasses

2 ounces ginger

4 ounces hops

6 gallons water boiled, strained & when blood warm, put in 1 quart yeast to make one barrel of beer. Let it work 1½ days & then bottle it.

LOAF CAKE

(recorded in Preeble, New York)

1 quart milk

5 pounds flour

3 pounds sugar

2 pounds butter

8 eggs

½ pint yeast

2 pounds raisins

1 ounce cloves

1 ounce cinnamon

1 nut meg

1 table spoon salt

1 quart milk, ⅓ of the butter and sugar to be mixed and left to rise: after which, add all the other ingrediants together with a gallon of brandy, or wine.

IRISH BUTTER

(Joel Hancock of Virgil, New York)

8 ounces salt

3 ounces loaf sugar

2 ounces salt petre

8 pounds of butter

(For 8 pound Jerkin Butter put it in a close tub for winter.)[21]

Also during this season, perhaps when holed up in a tavern or a tent during inclement weather, Randel took his rough draft of calculations of earth excavation, horse exertion, distances, descents, ascents, as well as many other additions and divisions, and crafted them into a publishable form that would earn Wright's enduring enmity.

By the end of the year, it appeared that Wright and the commissioners had quietly and privately settled on a plan. According to newspaper reports, the route was to go to Cohoes—and thus first to Troy—and thence to Albany. The canal would cross the Mohawk River two times and follow a northern path through the Mohawk Valley. The choice was certainly not what Randel had suggested, either in the *Albany Daily Advertiser* or in his communications with the commissioners and engineers.

On January 11, 1822, in apparent response to this news, a "Friend of the Canal" published a commentary in the *Albany Argus* about the eastern part of the route. The particulars of the western portion had all been published, the author said, so as "to attract the aid of scientific men, to unlock the repositories of useful, local, knowledge . . . The eastern section, on the contrary, has been managed as a close concern, withdrawn with retiring diffidence from the public eye." And so the Friend obligingly laid bare the issues for the eastern end, which was the scene of "two adverse conflicting powers, equally abhorrent of a straight line." The Mohawk River is "a broad unbridled and impetuous stream," and traversing it twice would be impetuous, ill-conceived. The chosen route made little sense, the Friend argued, even though Mr. Wright pronounced it "the most eligible, if practicable." More of the same appeared on February 5.

On February 15 the Friend took a new tack. The proposed route would not survive natural forces: snowmelt would bring the level of the Mohawk River dangerously high, dangerously quickly. There would be flooding, the volume of water would be too great for the canal. Debris and ice would destroy workmanship: "The embankments exposed to the pressure and motion of the water, will have become porous, leaky, demolished or dissolved." Repairs would finish just as the seasons of flood and ice were again to commence.[22]

On March 5, the *Argus* curbed the "Friend" series. The situation had changed, an editorial stated, because the commissioners had finally released their final plan and report. Notwithstanding the great respectability of the author of the Friend letters, the *Argus* maintained that on examination of the previously withheld details, the commissioners' current plan did make sense. The paper published an excerpt from the report. The Friend's insights had become superfluous.

Randel did not agree. A month later, as set out with little fanfare in his notebook, he published a forcefully written and argued seventy-two-page pamphlet, *Description of a Direct Route for the Erie Canal, at Its Eastern Termination: with Estimates of Its Expense and Comparative Advantages*. The author "regrets, sincerely, to be compelled to advance any thing, which, even by possible construction, may implicate the correctness of conduct of any public officer, and more especially officers vested with such high and important trusts as the Canal Commissioners or their agents; yet the immense importance of the subject requires that facts be stated without reserve, and thoroughly investigated."[23]

Randel arrayed in resplendent detail his evidence that the new route promoted by Wright and the commissioners was 14 miles longer and significantly more expensive than yet another, tunnel-less alternative route he had devised. (The tunnel having been discarded, he wrote, because Mr. Wright's concerns about the permanency of such a wall in the bed of the Mohawk River on the new route "are no doubt well founded," and thus "there can be no propriety in now urging the adoption of a tunnel for the eastern termination of an impracticable route for a canal." To wit, Mr. Wright was correct about the tunnel not working, but only because he had chosen an incorrect route.) His introduction stated that he had requested information from Wright, "in this expectation however, he has, unfortunately been disappointed; Mr. Wright having recently informed him that he has no leisure to bestow upon this subject." According to Randel, his proposed direct route would cost $714,855 versus $1,620,826 for the commissioners' new route. Randel's attention to figures was central to his view of himself as an engineer. As historian Daniel H. Calhoun notes, the ability to make an estimate that was realistic and that could direct whether the project should be done or not was what differentiated the "ordinary competent surveyor" and the engineer. The latter was thought to have that expertise—construction and cost were the engineer's responsibility."[24]

The document contained a chronology of events and communications as well as intricate details about elevation, water flow and supply, distance, terrain, labor, costs, and time. It also set forth intriguing ideas for methods. For example, Randel described how boats might remove excavated earth using "pyramidical troughs" running down the banks to an elevated crossbeam, like a miniature pier, under which boats could wait to collect the descending dirt. "These troughs will diverge as they extend upwards; and along their sides, the men employed in digging would be disposed, shoveling into them. Thus saving the expense of wheeling, and subjecting a whole face, or inclined cross section of the deep cut, to be worked at the same time." The idea was essentially arterial. A series of troughs could fan out across the canal bank and be placed next to the men who were digging. The contents of these many troughs would run down the bank and empty into a main trough. The main trough in turn would run into a boat, and the dirt could be floated away and dumped elsewhere. Men would have to shovel their dirt only a short distance, and it would simply slide away for removal, instead of needing to be carried away in a wheelbarrow, as was usually done.[25]

Randel may have escalated his Erie critique for a variety of reasons. Some Erie scholars have argued that he was in the pocket of Albany while Wright was in the pocket of Troy, that Randel was merely a front man and behind this pamphlet stood Albany powers, men such as Simeon DeWitt and Stephen Van Rensselaer. Randel's own notes support that conclusion. First, he had commented to James Geddes on July 16, 1821, that Albany would not give up easily. Second, the Albany mayor was among the men who hired him to do his survey. Third, he recorded that DeWitt gave him $30 to pay the printer for the pamphlet; although issued in his name, the pamphlet was not paid for by Randel. One writer has also argued that Randel was angry because his reports had been ignored and because despite all the excellent information he had put forth, a less "eligible" route had been chosen. It was probable too that Randel was furthering his own agenda and long-term business interests, establishing his prowess and exactitude as an engineer, his standing as an innovator and a man of science.[26]

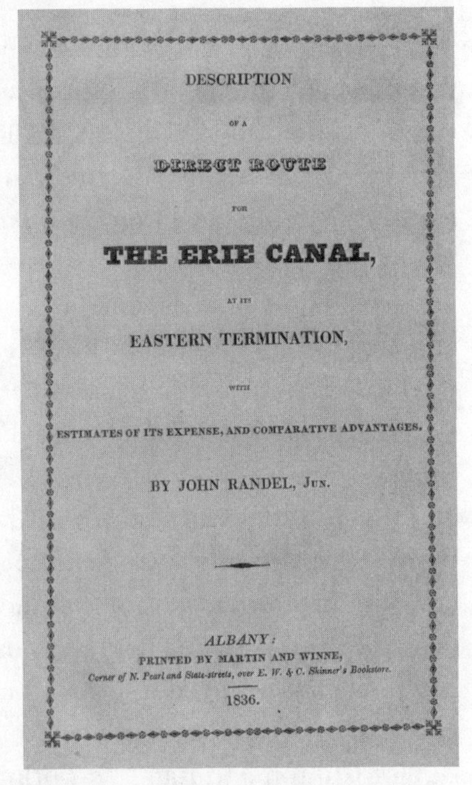
DESCRIPTION
OF A
DIRECT ROUTE
FOR
THE ERIE CANAL,
AT ITS
EASTERN TERMINATION,
WITH
ESTIMATES OF ITS EXPENSE, AND COMPARATIVE ADVANTAGES.
BY JOHN RANDEL, Jun.

ALBANY:
PRINTED BY MARTIN AND WINNE,
Corner of N. Pearl and State-streets, over E. W. & C. Skinner's Bookstore.
1836.

Description of a Direct Route by John Randel Jr., reprinted in 1836. Collection of the author.

Whatever the blend of reasons, Randel deferred to accuracy and correct calculations—lofty authorities. It is unlikely Randel would have made himself professionally vulnerable, for his mentor or for politics, without conviction and belief in his argument. Math contained a truth Randel thought was self-evident, above the fray—if he could just get Benjamin Wright and others to see the math, conflict would evaporate, truth would shine. They would build earth-removing troughs; they would follow the most fit and proper, the most eligible route.

But they did not. The Erie went to Cohoes and a feeder canal emptied into the Hudson opposite Troy, which sits on the east side of the river. From there the Erie traveled south to Albany.

After Randel published his pamphlet, his life resumed its normal rhythms. He sent grass seed—eastern clover, western red clover, and timothy—to Oneida. Matilda's uncle and aunt came for a short visit to Albany; relatives and friends stopped and stayed as they passed through on their way to Oneida or down to New Jersey. Randel had one of his assistants plant a garden of yellow potatoes, blue potatoes, and large potatoes, despite "ground very dry and dusty." In late May 1822 he set out for Oneida, and then traveled on to Syracuse to survey for DeWitt.[27]

Unfortunately for Randel, the day he reached Syracuse, three canal commissioners arrived by boat, had tea, and continued on to Salina to review work with Benjamin Wright and Canvass White. Over an early breakfast the following day, Randel bumped into Wright. His account of that meeting is one of the most cryptic and confusing of his field book entries. He noted that he asked if Wright had received a letter from him. Randel then launched into a many-page description of past events involving governor DeWitt Clinton and Randel's work on the route and a sealed letter with specifications from Randel in it. He said he gave the letter to Wright, who read it and then told Randel he should make clear his authority on certain subjects. "For example you mentioned how boats are loaded at Spanish River. Now I don't know anything about it, nor where to find any account of it," Randel quoted Wright as saying. "I told him my father was taken prisoner by the British in the Revolutionary War and arrived to that River where he observed this process and that I did not know that there was any account of it published." Spanish River, now Sydney, was a town on Cape Breton in Nova Scotia and the site of a major Revolutionary War battle in 1781. While a prisoner, Randel Sr. had apparently observed something similar to the earth-removing troughs his son later proposed in his Erie pamphlet.[28]

Randel concluded the entry with what was, if accurate, Wright's indictment of himself. "He said I had no right to expect him to answer such specifications as he could not do it without incriminating himself which he would not do for any resolution of the Commissioners. It could not be expected of him that he should acknowledge he had done wrong." Randel's entry is silent on how breakfast finished.[29]

Randel knew by the end of the year that he had badly erred. Engaged as he remained with Wright and with trying to set the record straight—perhaps only as a form of self-justification—he could see that the Erie stance had hurt him. "The stand I have been compelled to take in opposition to the Engineers of the Erie Canal makes it thoroughly presumptive with those who are unacquainted with the subject in dispute that I am deficient in skill as an engineer. It is therefore of the first importance to me that I be placed in a situation where I may have the opportunity to act," he wrote to DeWitt on Christmas. "To recover my former standing will now require an effort which unaided by you I fear will fail for as I have been compelled to oppose the egregious errors and waste committed by the Engineers on our canal they in return are compelled in self defense to contend that I am ignorant of the subject upon which I have written."[30]

In the immediate short term, this meant that Randel hoped DeWitt and other patrons, including Governor Clinton, New York State chief justice John T. Lansing (whose land Randel had surveyed in his youth), and Stephen Van Rensselaer, would provide references for an application. Randel had seen an advertisement for a civil engineer for public works in Virginia, and in December 1822 he wrote to inquire what kinds of public works the job entailed and at what salary. The deadline was January 31, and although DeWitt and Clinton did send letters of recommendation, Randel didn't apply. He missed the deadline. As it turned out, he had already been asked to consult on yet another canal. His involvement with the Delaware & Raritan Canal and the Erie Canal, despite his frustrations and anxiety about both, had finally paid off and brought him to the attention of others bound up in the country's canal fervor.[31]

We Regret Extremely Your Domestic Afflictions

While New Jersey and New York politicians and businessmen had been advocating for the Delaware & Raritan and the Erie, respectively, a tristate alliance of politicians and businessmen was similarly engaged not far to the south. The idea for a short canal cutting east-west across the neck of the peninsula between Chesapeake Bay and Delaware Bay predated Penn's for the Delaware & Raritan. Augustine Herman, a surveyor for Lord Baltimore and one of Maryland's first settlers, had proposed it as far back as 1661. Ralph Gray recounts the story of this short, significant canal in wonderful detail in *The National Waterway*, one of the few books to explore Randel's history and character at some length. In brief: Philadelphia businessman Thomas Gilpin initiated surveys in the late eighteenth century and in 1803 assembled a board for the canal company. Two potential routes were identified: an upper route, running from Elk River, near the town of Elkton in Maryland, to Delaware Bay, either at the town of New Castle or via Christina River; or a more southern route that would run from Back Creek, Maryland, to St. Georges Creek, which emptied into Delaware Bay. In 1803 the upper route was chosen, and work began the following year under the direction of Benjamin H. Latrobe, whose testimony would accompany Secretary of the Treasury Gallatin's report. The next year it ceased for reasons of politics, topography, labor, and capital.[32]

Several years later, after the War of 1812, Mathew Carey's exhortations revived the project. Promoters such as Carey understood that such a canal, although less than 20 miles long, would shorten travel between Philadelphia and Baltimore by several hundred miles—300, in fact. It would also provide a crucial link in an inland waterway that could protect vessels from dangerous sea travel, an idea central to Gallatin's report. In March 1822 an engineer named William Strickland was engaged to do yet another survey and review the earlier choice of the upper route. A few months later Strickland said he supported the original plan, with a slight modification. The company's board of directors, however, was not convinced.

That same year the Chesapeake & Delaware Canal Company directors wrote to New York governor DeWitt Clinton, asking him to recommend engineers "of competent talents & experience." Before the end of the year, the board hired Wright. In December 1822 it also hired Randel to review all previous surveys and reports and to locate the "most eligible line," for which he was to receive $200. That Randel and Wright were working together seems not to have daunted Randel. As of March 1823, his records suggest no ongoing tension with the judge. Indeed, quite the contrary. He noted to the president of the company that Wright wanted his help as they examined the various proposed routes. Randel also wrote to Turner Cormac, another company employee, that Wright had approved his cost estimates and his dirt-directing troughs, noting, "I have taken much pains to explain to him the position of the troughs and boats . . . I find that he had never before, as he now acknowledges to me, fully understood how they were to be worked. Now he understands them, and sees everything plain and easy he cannot imagine any reason why they should not exactly answer this purpose.[33]

"It may be a matter of surprise to you, how all this has come to pass," Randel went on to Cormac. "We met accidentally in one of the public offices, he expressed a desire to see my maps etc of this route and I invited him to my house. Here I laid before him my report and all my maps, plans, estimates etc answered all his inquiries and endeavored to give him . . . a complete knowledge of the country. Although Mr. Wright has been my opponent, but not so I believe any more, I have never ceased to consider him a man of sound judgment." Randel was naïve. Wright had nothing in the vicinity of a similar opinion regarding him.[34]

Randel's recommendation about the Chesapeake & Delaware's best route challenged Strickland's and Latrobe's selection of the upper route. Unsurprisingly, Randel championed the more technically challenging southern route, which would require what came to be called "the deep cut," a profound slice through the earth that would render the canal sea level, permitting an influx of ocean water. The proposal solved the problem of ample water to fill the canal, something the company and all the surveyors had worried about. Although filling the canal high enough to carry boats would no longer be a challenge on Randel's route, there were other obstacles. The route demanded greater excavation, particularly across one 80-foot ridge. Randel "is full of the throughcut," reported the company secretary, Henry D. Gilpin, to his father, Joshua Gilpin (son of founding board member Thomas Gilpin). The deep cut Randel advocated was, according to a later assessment, considered "one of the greatest works of ingenuity and skill in the world." It also proved more expensive, because the soil was so wet and continually sloughed off what were supposed to be firm embankments.[35]

At the time Randel championed the deep cut, the idea was generally unpopular. "There is a powerful party opposed to Randel consisting of all Strickland's friends," Gilpin noted. Nevertheless, the company kept its options open and requested that the War Department review and assess Strickland's and Randel's proposals. At the same time it hired Randel's younger brother William to test the quality of the soil. William arrived on May 14, 1823, and set out to "bore along the line his brother has laid out." Wright arrived after a time; Gilpin found him "a very nice man." Although Wright's mind was not made up, he naturally favored the path Randel had rejected; "far from having any bias against the upper route, he is evidently inclined to it, if he can get water which he looks upon as the great impediment," Gilpin wrote. June and July were filled with borings and soil testing. The summer also brought a visit from the U.S. Army's Board of Engineers.

(Engineers had been part of the military since the Revolutionary War, and Thomas Jefferson established a permanent board in 1802, now called the Army Corp of Engineers.)[36]

Traveling from Albany to Delaware and Maryland, a journey of several days, was painful for Randel during these months. Matilda had become quite ill. In March, Randel wrote to his friend John Telfair that "Mrs. Randel has dropsy in her bowels accompanied by a tumor external of her abdomen. She can walk about the house but has lost flesh so as to make her bones project. As to the tumor three doctors differ in opinion. One not knowing what it is, but certain it is no tumor. The other two being certain of it being a tumor. All of them are skillful men. She therefore in this state of doubt only takes medicine for the Dropsy. Father and mother are in bad health. My anxiety and fears on her account you may imagine better than I can describe." Matilda moved into her parents' house in Bloomfield, where Randel stayed by her side. She died on August 11.[37]

From that moment on, Randel's life had a different quality. His life with Matilda seems to have been a moored one, family-centered, happy. After her death, Randel was often embroiled in some controversy, some court case; the records that endure suggest periods of imbalance and poor judgment, even paranoia.

After Matilda's death, Randel briefly stopped working. Henry Gilpin understood. "Be assured Sir that your domestic afflictions must render needless any apology for the little delay that has occurred in sending on those instruments for completing the boring," he wrote. Randel did get back to work on August 28. Then October brought the death of his father. And Randel himself apparently became—or had already been—very ill. "We regret extremely your domestic afflictions, not only as they obstruct the prosecution of your labor on our behalf but as they have been attended with the production of so much suffering to yourself," Gilpin wrote. A decision about the route had to be postponed until the army engineers could meet with Randel to review the questions they had about the lower route. They could not do that until later that winter because of his poor health.[38]

A FEW MONTHS LATER, at the end of January 1824, Randel received some good news: the Army and the company had selected his route and were to hire him as an engineer. But disappointment arrived hand-in-hand. Randel was not to be in charge. The board had appointed Wright chief engineer of the entire project, and Randel was to report to him. His assignment was to construct the eastern half, the site of his deep cut and, as on the Erie, the most challenging portion of the canal. Not only did it contain the highest elevation (the aforementioned 80-foot ridge), but it was thick with marshland and boggy ground near the Delaware Bay terminus. On March 26, Randel signed his contract. It seems safe to assume that neither he nor Wright and the canal company directors ever imagined the legal scrutiny that contract would soon undergo. Randel was to have four years to finish his section of canal, which was to be 10 feet deep and 66 feet wide at the surface, narrowing to 36 feet wide at the bottom. He placed advertisements for workers in newspapers in Massachusetts, Connecticut, New Jersey, Pennsylvania, New York, Maryland, and Delaware. And, ever proud and self-promoting, Randel advertised the fact that his route had been selected over the others, the directors having "unanimously adopted the *Deep Cut Route*, as projected, laid out and recommended by the subscriber."[39]

Work began in April. Randel devised a tramway, one of the country's first, to remove excavated earth. (Perhaps the pyramidical troughs were not expedient on that section, although it is hard to imagine that he would not have insisted on building and testing them.) Hundreds of men were employed, and Randel was instructed to work through the winter, forcing his men to labor in frigid water. As was typical, Randel received a salary, out of which he was to pay his crew and other expenses. After he established his subcontracts with laborers, though, Wright—with the approval of the board of directors—reduced his pay. Bound by his subcontracts but now earning less than he had been promised, Randel started losing money. In addition, worried about the schedule, he bought some costly machinery for underwater work. He remained ill, pushing himself physically despite his aliments, and pushing himself further and further into debt.

Wright well knew Randel's situation but was not disposed to intervene. One of the canal company board members resigned, however, protesting the mistreatment of Randel and saying that he could not longer "conscientiously" continue on the board. Paul Beck Jr. noted that Randel's outlay for men's wages was five times greater than what he himself earned: "I resigned because from the acts of the board toward Randel, it was manifest to me that his ruin was inevitable; and I did not choose to be accessory to it . . . I specify the allowance on one occasion of $600 for work amount to $3000; I speak, however, of the tendency of their whole conduct. I believed that the board was governed by Wright, and that he had determined to ruin Randel."[40]

During this period Wright wrote a letter to John B. Jervis, a civil engineer known for his canal work and his later work on New York City's Croton Aqueduct, who had apprenticed with Wright, airing his intense dislike of Randel.

> I go on here as I expected, *in hot water with my worthy friend Mr. R*, who always moves by high purpose and will probably as all other high purpose steam engines eventually burst. All that I am afraid of is that I am too near him and some of his steam may splatter on me; as I find he is much disposed to throw his steam in all directions—altho his own wish would be to direct it toward me—as he considers me as standing in the way between him and "everlasting fame." I hope he will sleep better when I am further off from him than he now does—for in a few words I think him the most complete *hypocritical lying nincompoop* (and I might say scoundrel if it was a Gentlemanly

> word) that I ever knew except it is his Brother and Old Father Putnam and they are both here and excellent aids they make for him.[41]

As chief engineer, Wright was in a position to indulge his animosity, to make life intensely difficult for Randel. Randel's contract with the canal company gave Wright the power to assess his work, and if it was to be found lacking, according to Wright, and Wright alone, the company could annul the contract. Despite reduced wages, ongoing illness, and mounting debt, Randel worked continuously through the winter of 1824 and into the summer of 1825, when records show he had 514 men working for him and 154 teams of horses. It was miserable work. "I did not think that Randel's workmen managed well; they were too much in mud and water," commented an observer.[42]

Randel's unrelenting pace, despite great cost to himself, was not unusual. His Manhattan and upstate New York fieldwork had shown the same quality. Wherever he was, he pushed himself and his crew to work hard, quickly, and with exactitude. But as he cut through spongy mosquito-infested marshes in Delaware, he must have realized that Wright was sabotaging him and decided that excellent performance would be his defense. He must have also reasoned that a success on the deep cut would elevate his professional reputation, even if he was not engineer in chief.

Once again Randel was wrong. Wright routinely visited him in the field to assess his progress and to certify the number of cubic yards of the canal that had been dug. Despite the ample progress others saw—Randel had finished close to half of his section within a year, well ahead of schedule—Wright reported to the Chesapeake & Delaware Canal Company directors on July 30 that Randel had neglected his duties. It later came out as testimony that Wright lied and instructed his assistant to lie. "I did hear Benjamin Wright give directions to Henry Wright to make short and false estimates and certificates of Randel's work. He said he would ruin Randel's credit and break him up," noted an observer named Dr. Gemmel. At least one member of the canal company board was a party to the dissembling. Henry Wright received a letter from a director at the company, "approving of his conduct in relation to Randel, and requesting him to persevere in it. Wright showed me the letter," said Dr. Gemmel, "and immediately burnt it."[43]

Benjamin Wright and his men had support on the board of directors, but even so, there was much talk about what to do concerning Randel. The directors met in August to discuss Wright's "certificate," or report, on Randel's neglect of duty. Whispers about that meeting soon made their way to the work zone. "I heard various rumors at the canal that the contract was about to be taken from Randel; there was quite a riot, and the work suffered much hindrance in consequence of these rumors," said engineer George W. Smith. Smith went to talk with the canal company directors, who assured him the rumors were unfounded. "I returned to the canal and endeavored to quiet the apprehensions of the workmen and others . . . I am confident that John Randel did not know of the certificate until it was officially communicated to him."[44]

On September 10, Randel was finally informed that Wright's certificate showed neglect of duty. On September 12 he wrote to the canal company requesting the specific details of his failing—presumably to counter with his own tallies of cubic yards dug, miles finished. The board refused his request. On September 19 the directors relented and met with Randel, who asked "for time to prepare his defense." He was given ten days. Randel was, again or still, very ill. His friend Smith tried to intervene on his behalf. "I stated that Randel was ill, had been cupped on his head and neck; that he was then attacked with coma," said Smith, who was a well-traveled man of science, who greatly admired Randel, and who noted that one of Wright's canal choices "has been the subject of much ridicule among scientific men; it is perfectly absurd." Smith sought to read Randel's defense to the company in his stead. "They still refused to let me appear for him. I remarked that they treated him with neither justice nor common decency." Smith found the company's refusal "unjust and cruel."[45]

The following evening the board met again. "I had hoped this day instead of writing to go down by the boat," Gilpin wrote to his father on October 1. "But last night after great discussion the Board discharged Randel, and this of course puts us in a good deal of bustle—and prevents my leaving town at present."[46]

All the discussions about Randel during August and September had rankled Wright no end. "I had a full belief that all my troubles would be at an end on my arrival here, but I find it is not so. This J.R. is so full of his lies and schemes of trouble that I have a new fence before me and as much correspondence as a Minister of Foreign Affairs of any nation," Wright wrote on September 11 to Jervis. Wright contemplated leaving the company around this time but was apparently begged to stay. "It is really too bad to be so placed—but what shall be done? These Gentlemen are so friendly and totally unable to get on with their difficulties with R without me to protect them that they will not listen a word to my quitting them & say that my reputation is connected with theirs and we must go together." He noted in another letter, "the drama thickens apace. And I think the Board will be obliged to come out & say something—I had a piece prepared for the paper and was determined to publish it. But the Board said & begged I would not." Wright must have hoped he would be finished with his nemesis once Randel was fired. But their feud just changed venues.[47]

Nothing but Sickness and Sorrow

For Randel, the dismissal was shocking. "GENTLEMEN, I have received this morning, with equal surprise and affliction, your resolution of the 30th Sept.," he wrote on October 1. "I hope it is no satisfaction to you or to any one else to know, that having sacrificed my health by unwearied exertion and exposure in your service, *you have now employed your combined power and influence to ruin my fortune and my reputation,* and to *leave me as a reward for all I have done and suffered, nothing but sickness and sorrow, and perhaps a broken heart.*"[48]

A lawyer later characterized Randel's shock in less emotional but equally descriptive terms. "Randel must have been amazed to learn that he had abandoned the work in July when he knew that all August and September he had been delving in the mud, in the midst of autumnal fevers, paying five hundred men their daily wages, exhausting his funds and draining his friends; and submitting himself twice a month to the inspection of the 'committee of works,' who, armed with champaign and Cook's pills, ventured in this unhealthy region to see how fast a man could *complete* a work that he had *abandoned*. Surely he must have been in a trance all this time, or the abandonment is false and its operation fraudulent."[49]

Mathew Carey, one of the men most instrumental to the canal company's creation, found the treatment of Randel unconscionable. Carey plunged into the fray to defend Randel, alienating many of his colleagues, business associates, and social acquaintances. His first support of Randel emerged in a formal protest printed just a week after the dismissal, on October 8, 1825. Carey hoped Chesapeake & Delaware stockholders ("such of them as are friends of justice—as abhor the idea of crushing an unprotected individual by the exercise of uncontrolled power") would act. In his protest, he outlined a chronology through which, in his view, Wright and the board's "spirit of hostility" toward Randel was revealed. He noted that the fees to be paid Randel by the company had been reduced, that Randel had lost money and had been sustained by the "aid of his friends and his own private fortune." (What private fortune that could have been in 1825, a few years after Randel had complained about his inability to pay for his house in Albany and after numerous other complaints of financial struggle, is not clear. It may well have been that there was a discrepancy between how Randel was perceived or presented himself and the reality of his finances.)[50]

Carey went on to chronicle other abuses, including the endangerment of Randel and his workers' health by making them toil unnecessarily on a frozen marsh. He laid out contradictions between Wright's account and that of the board. About Wright he had nothing good to say: "Because Judge Wright, the accuser in this case, was an infinitely more suitable object of accusation himself, for neglect of duty; as, while he was in the receipt of a large salary from the board, he was absent the chief part of the spring, and almost the whole summer, surveying sites for canals in other states." Carey noted that Wright thus avoided the "sickly season," while Randel did not.[51]

Carey's criticisms were well founded. Wright's fear of working in unhealthful conditions was transparent in his correspondence to Jervis. And Wright had often been absent on the Erie as well. Indeed, he had opponents among the Erie commissioners, several of whom did not approve of him or his methods. It must have upset Randel deeply to see Wright—a less attentive, less inventive engineer—reaping acclaim for the Erie Canal, which opened to great fanfare in October 1825, just as Randel's work on the Chesapeake & Delaware was being unjustly criticized. And it must have upset him to see the Chesapeake & Delaware Canal Company, which had chosen his route, so utterly turn their backs on him. Repeatedly and unsuccessfully, he sought to meet with the directors and to obtain a written list of his offenses and failings.[52]

Carey, however, was able to provoke a response. On October 13, one stockholder wrote that he rejected Carey's version of the truth: "In management, Mr. Randel is notoriously deficient, as can be proved by the acknowledgments, at many different times, of George Gillaspey, Esq., a member of the board—and he never had an hour's experience in canalling until he commenced this work; though he published a book in New York, upon the subject, that did not add to his credit or importance." The author went on to say that it was Wright who got Randel the contract: "In this the Judge was culpable; to recommend a man without experience, inflated with vanity, and extremely visionary." He denounced Randel's abilities and his nepotism in hiring his brother William. "Extremely visionary" was, in this context, an insult. Canal work was perceived as practical work undertaken by practical men for practical ends. This stockholder, and others of that time, deemed a visionary approach excessive, more individualistic and indulgent than the situation required. (A newspaper article about canals and rails captures the contemporary meaning: "This is no visionary, or impracticable scheme, got up like some bubbles of the day, to delude the unwary and inexperienced.")[53]

This response only deepened Carey's resolve. His first protest occupied a single printed page and went into select details. His next publications, called an "appeal" and a "last appeal," were longer and published in several editions. Then he issued the twenty-three-page *Exhibit of the shocking oppression and injustice suffered for sixteen months by John Randel, Jun. Esq.* Carey said that when he had written his protest he had possessed "an imperfect view of the affair, and was by no means aware of the extent of the injustice and oppression which Mr. Randel had struggled with." Now that he had a "view of the whole ground," he hoped to show that "the United States has scarcely ever witnessed a harder or more cruel case." Carey called for a juried trial. He listed sixty-six facts that proved the innocence of Randel and exposed the company's shoddy behavior.[54]

Carey stated that he had no conflict of interest: "With Mr. Randel I have no concern whatever, to the amount of a cent. My acquaintance with him has been very slight; is of quite recent date; and will probably terminate with the settlement of this affair, as there is no congeniality in our habits or pursuits. Of course I have no private or personal motives to stimulate me to the course I pursue."[55]

Carey's motivations were likely more complex. He had published an attack on the "horrible oppression" of Catholics (his faith) in his native Ireland, and as a consequence had needed to flee and hide for a time in Paris, where he met Benjamin Franklin. He retuned to Ireland, only to have to flee again because of a libel charge, which came on the heels of jail time because of his anti-English writings. He boarded a ship—so his story goes—dressed as a woman ("and must have cut a very gawkey figure") and headed to Philadelphia, where he established his printing company. Edgar Allan Poe described Carey as having a "hatred of oppression" in his 1836 review of Carey's autobiography. Carey spoke out on many issues throughout his life, including religious freedom, public charity, and unequal pay for women. "But can they withhold relief from her who comes in her desolation and weakness—*woman, who, by the law of her being, is excluded from paths in which coarser men may make a livelihood*; and, by the custom of society, is OBLIGED TO ACCEPT LESS THAN HALF OF WHAT THE MOST STUPID OF THE OTHER SEX CAN EARN, *as a compensation for her unremitted toil*," he wrote in one of his essays. His rhetorical flourishes were not limited to his defense of Randel.[56]

Mathew Carey by John Neagle. Courtesy of the Library Company of Philadelphia.

James N. Green of the Library Company of Philadelphia, an expert on Carey, thinks Carey's involvement with Randel's case had less to do with righteousness and more to do with canal advocacy, "which takes over his life between 1825 and 1830, until he gives up in disgust." Beginning in 1821, Carey had first pushed for the revival of the Chesapeake & Delaware

Canal, and in 1825 had pushed for a canal over the mountains of Pennsylvania. "Because he was such a good writer and such a stubborn person, he convinced otherwise intelligent citizens of Philadelphia to support it. [Randel] ties into his crusade," Green says. "He probably doesn't care that much about Randel. He is appealing to the sense of justice to try to get Randel out of this [controversy] and get back to building the canal." If true, Carey's strategy backfired. The lawsuit delayed work on the canal. And Carey alienated other engineers as well as some important businessmen.

The passionate defense by Philadelphia's "Fiery Irishman" even may have convinced Randel and his entourage that the case was strong enough to pursue. "Judge Geddes and Simeon DeWitt called at 7, with J. Randal. Had a long conversation with them on Randel's affairs, which I explained to them. They are perfectly of accord with me," Carey wrote in his diary. (The inclusion of Geddes at that meeting is notable, because of Geddes's involvement with the Erie Canal and his disapproval, as expressed in Randel's field books, of Randel's outspoken critique of Wright a few years earlier. It suggests that the company's treatment of Randel was indeed perceived by many as egregious.) Although many of Randel's patrons and associates may have supported Carey's strategy, Randel's lawyers did not always heed his advice. "I am sorry I cannot do what you wish," John Sergeant wrote to Carey. (Sergeant initially represented Randel.) "I have, as I hope you know, the greatest respect for your purposes and motives upon this and upon all occasions, believing them to be generous and liberal. But since the conversation in my office, you have been aware that the counsel of Mr. Randel did not agree with you as to the best mode of obtaining relief for him. In my relation to Mr. R., I must be governed by his wishes and views, aided by the other professional gentlemen who have the charge of his case, and all of whom I believe agree in opinion as to the course to be pursued." There is no record of Carey's response.[57]

Randel's collision with the company was in some regards unexceptional. Many engineers of that era found themselves in similar situations. Historian Daniel H. Calhoun notes that Loammi Baldwin, a contemporary of Randel's also dismissed by a canal company, felt frequent frustration because his employers had no real understanding of the work: "All this *splendid* display of wealth and science was made under *watchful* committee, and subcommittees of Boards of managers, without the aid of engineers. And can you wonder at the consequences." Baldwin, whose independent suffer-no-fools personality resembled Randel's, also noted that "the circumstances under which the engineer is placed, are too confined and perplexed, for him to perform his duty with facility and despatch . . . In short, the engineer has a great deal of labor and responsibility, but no independent power." Charles Ellet, an engineer who worked on canals and who sought to establish suspension bridges in the United States, had similar experiences. Many years after Randel's troubles with the Chesapeake & Delaware Canal Company and Benjamin Wright, Ellet was dismissed by the James River and Kanawha Company for, in essence, disagreeing with Wright. Ellet, like Randel, seemed to prefer working as an independent; he too was driven by visions of innovation. As Calhoun notes, such individual temperament increasingly lost out against the "company man" as the nineteenth century progressed.[58]

Wright was a prototypical company man. He was likable, as Henry Gilpin noted. By contrast, Ralph Gray describes Randel as volatile and "quarrelsome by nature." This had not always been true. When working in Simeon DeWitt's upstate office, Randel had been agreeable; it is unlikely that DeWitt would have so favored him had he been otherwise. And although he irritated the Common Council of the City of New York by requesting more money, he does not seem to have fought extensively with the aldermen either. His "quarrelsome" nature may have originated with professional insecurity, which grew after he left Manhattan.

It was a good strategy in some instances. Quarreling about the route of the Erie had brought Randel into the public eye, perhaps even secured him the Chesapeake & Delaware Canal contract. Quarreling about the miserable situation in Delaware was a sane, justifiable response. But after the events of 1825, Randel's combativeness seems to have developed an embittered quality and to have become more deeply rooted. Thereafter, quarrelsomeness seems to have defined his personality everywhere he went.[59]

In the early days of the conflict, a negotiator named Nathan Bunker went back and forth between Randel and the Chesapeake & Delaware Canal Company, seeking to reach some compromise. "I believe Jno. Randel, jr., and the company had no intercourse. I understood they could have none. I had myself an interview with J. Randel, at my own fire-side, very early in the controversy, (in the stage of it that I took part in,) in which I had hoped some good might result. When I conversed with the canal company the next day, I found they were extremely wide apart in their feelings," Bunker later described. "I found there was an exceedingly hostile feeling between John Randel and the company. When I discussed with either party the matter in controversy, I found a very angry feeling between the parties, and no desire to conciliate."[60]

After consulting with lawyers and collecting materials, Randel filed a suit against the company, charging breach of contract. Philadelphia's district court registered the suit on January 1, 1826, but ruled that the case instead should be brought in one of the states where canal tolls were collected—Maryland or Delaware. Randel and his legal team then brought the case to the Court of Common Pleas in New Castle, Delaware, on June 18, 1828. On a matter of jurisdiction due to a change in state law, the case was then transferred to the Delaware Superior Court in 1832 and was finally heard there in 1833, eight years after Randel's dismissal. During those years Randel went into even greater debt, postponing payment of bills, lawyers, and assistants until the case was settled, banking on the belief that the suit would be resolved in his favor. He had as many as four clerks copying papers, going over records, contacting potential witnesses. "He retained a great many professional gentlemen, all that he could reach," according to one account.[61]

Randel's legal team was led by John M. Clayton, a significant political figure in Delaware as well as in Washington, D.C., and "fully master of every weapon of argument and eloquence," according to contemporaries. Randel's case against the company was not the only one Clayton was working on. In 1831 the Delaware Court of Common Pleas heard a case of trespass that, according to a press report, "has excited an unusual degree of interest." Landowners along the canal had sued the company for flooding their land, "by which they have lost use of the marsh, and the health of the neighborhood, it is said, is materially affected." The company had shown "unpardonable indifference." Clayton and George Read Jr., another member of Randel's legal team, prevailed. Their clients won $6,000 in damages.[62]

That same year Clayton and his team also brought a case against Benjamin Wright, *John Randel Jr. vs. Benjamin Wright*. Randel's lawyers contended that Wright had been given a position over Randel in which he was to act impartially, but that he instead "maliciously" sought to harass and embarrass his "rival engineer" and to turn the company against him. The lawsuit placed damages at $100,000. The Court of Common Pleas declared a nonsuit on the grounds that the case was really between Randel and the company. Randel and his legal team appealed, unsuccessfully. "At the June Term, 1832, that court affirmed the decision [of the lower court]. The case was not much argued, if at all, in the Court of Appeals."[63]

The case between Randel and the company, however, was much argued. "The troublesome nature of the controversy may be inferred from the facts, that the counsel for the canal company filed sixty-two pleas, to each of which there was a replication or answer. The whole of these were afterwards withdrawn: the record broken up: new counts added to the declaration: twenty-nine new pleas and demurrers filed, to each of which there was a replication or a joinder in demurrer, as the case might require, all of which were drawn out at full length," the U.S. Supreme Court later noted. (Demurrers argue for dismissal; the documents may admit that the facts of the case are true but nevertheless maintain that there is no ground for a lawsuit.) The endless back-and-forth filings drew out the case. When it reached Delaware Superior Court in the spring of 1833, the judge, Samuel M. Harrington, declared that there had been enough. He ruled on the demurrers, finding for Randel: "And let these judgments stand, as both sides have heretofore been allowed to amend repeatedly; and they have come down to the argument and to judgment on these demurrers with their eyes open. There will never be an end to this cause if the parties are to demur when they please, and amend as often as the demurrers shall be ruled against them." Harrington then set a date for the case to be heard the following term.[64]

Although Harrington had been forceful, Randel and the company did not stop generating demurrers in the winter of 1833. Beginning on November 29, Harrington dispatched their most recent motions. "The court was engaged a week in hearing the argument of some new demurrers which had been filed by the defendants since the proceeding term," reported the *National Gazette*. Finally, on December 9, the jury was empaneled. The infamous case, *John Randel, Jun'r. vs the President, Directors and Company of the Chesapeake and Delaware Canal,* commenced.[65]

According to a report in the *Baltimore Gazette and Daily Advertiser*, "the testimony was voluminous . . . and (with the exceptions and arguments arising on it) occupied a week in laying it before the jury. The arguments of all the lawyers took more than a fortnight." Clayton was a "tall, commanding, thoroughly well-developed figure," according to a memoir by a colleague. And he was deeply absorbed by the case. "I shall never forget the labor, as an amanuensis, he required me to perform at the time he was pleading to issue the great Randel case," the colleague wrote. "Most of the pleadings in the case were dictated by him, without any book before him, as he walked the floor of his private office . . . As in the Randel case, so in other cases. His whole soul was, as it were, given up to them, where there was to be contest. He would think, or talk of nothing else; you must listen to him about his case, or question, or leave him."[66]

The lead attorney for the company was James A. Bayard, with whom Randel later became entangled over some land transactions. Bayard and his team sought to prove, with various measures of cubic yardage, that Randel had not fulfilled his duty. Bayard's principal strategy was to object to the introduction of evidence—everything from the original contract to testimony, including that of Paul Beck Jr., the canal company board member who had resigned, and that of Randel's bother. William had been deposed in Albany in October 1829, while sick with a "pulmonary consumption" that left him too weak to write or to read aloud. The clerk appointed by the company to question William also became sick during the deposition; he "was seized with mania a potu"—what we would call delirium tremens resulting from drunkenness—"and his friends carried him off." Another clerk finished the interview. William died a year later, in October 1830. Now, three years after his death, the canal company was objecting to his evidence. Clayton was incensed: "William S. Randel's deposition has been returned near three years; opened, published, and a copy actually taken by the defendants. During the life of the witness no objection was made to the execution of this commission, and no exceptions have been filed since." After copious objections by Bayard et al., Harrington ruled for admission. "If ever there was a case in which a prepared deposition would be allowed, this is such a case," he noted. "The witness was languishing in a dreadful disease, which would certainly prove fatal, and whose violence would be greatly excited by conversation." In short, the company owed it to the witness. He had given his life to comply with their wishes and make the deposition.[67]

In his summation, Bayard had to address the notable absence of a witness for the defense. As he put it: "There is another matter—the Gentleman has said that Benjamin Wright has been kicked out of the cause or like a trembling coward forsook his master. It is true we have been deprived of his testimony. We took extraordinary pains to summon him as a witness. We had him here for two weeks at a great expense it then became necessary for him to go to New York and on his return found that the Court had adjourned. He then went back to New York and we daily expected his presence here until the close of the testimony and on account of his absence many things will remain in doubt which could fully have been explained." Wright never took the stand.[68]

On January 21, 1834, Harrington charged the jury. "The period has at length arrived in the progress of this cause when you are to become the chief actors in it," he said. "We are trying, gentlemen, an action of *covenant*: an issue of *breach* of covenant: an inquiry into *loss* and *damage* arising from breach of covenant: and a claim of *compensation* in damages *for* breach of covenant." The jury deliberated for four days and returned with a stunning victory for Randel: an award of $229,535.79 (including damages and costs). In today's terms, the settlement would be more than $6 million—then an unprecedented amount.[69]

Many newspapers relayed the "great verdict," italicizing the stunning figure. "This laborious and important cause was concluded on Saturday . . . Thus has terminated the most arduous trial, with the heaviest verdict, sounding in damages, that, we believe, has ever occurred in this country," reported the *Delaware State Journal*. The size of the award suggests many interpretations, among them: Clayton had presented a mesmerizing and compelling argument, much more so than that of the counsel for defense; Randel was likable; Randel's position was one the jurors could identify with; the canal company was disliked; the evidence of mistreatment and breach of contract was irrefutable. Whatever went through the minds of the jurors, they, to a man, felt the company had treated Randel shamefully and should suffer the consequences.[70]

Randel must have been euphoric. His decade of suffering was apparently over. He had been powerfully vindicated, exonerated. He could repay his debts, carry his head high, live well. His professional conduct had been upheld by the court. But the Chesapeake & Delaware Canal Company flouted the ruling and set out to ensure that Randel would never collect. Exceedingly hostile feelings persisted.

Whose Integrity and Correctness

The court case consumed much of Randel's attention between 1825 and 1834, and at the outset must have made him despair about his professional future. Who would want a man accused of dereliction of duty, despised by one of the most powerful and famous engineers in the country (Wright had built the Erie Canal!), a man who would take on his employers in such a public way? Nevertheless, Randel found ample employment as the case made its way through the legal system. His reputation for accuracy as a surveyor was untarnished, and so, it seemed, was his reputation as an engineer. He was hired for several jobs, including a canal survey in Pennsylvania and work in upstate New York, and appointed chief engineer for one of the first American railroads—direct competition for the Chesapeake & Delaware Canal, which must have brought him no small delight.

Randel also married again. He had fame, or at least a high public profile, and continued employment despite the lawsuit, so it is unsurprising he would remain a promising match. Letitia Massey was the daughter of Sarah née Strong and John Massey, a Philadelphia shipping merchant, who probably moved in the same business circles as the canal directors and Carey. Letitia had attended the Bethlehem Female Seminary (now Moravian College), the first school for women in America. Letitia, thirty-one, and Randel, thirty-nine, were married in June 1827 by the Reverend Benjamin Allen, rector of St. Paul's Church in Philadelphia. Allen had known Randel during his marriage to Matilda and approved of him. The reverend had written a letter of support that Carey published in his extensive pamphlet (along with letters from Simeon DeWitt, DeWitt Clinton, Stephen Van Rensselaer, and an upstate New York judge, Isaac Davies). "Having known Mr. John Randel, Jun. *fourteen years*, and being requested by a friend of his, to state my impressions concerning his general character, it gives me pleasure to say, that *I never knew any man of whose integrity and correctness I have been in the habit of entertaining a higher opinion*," Allen wrote. (A few of the letters Carey published predate the controversy and appear to have been the letters of recommendation Randel had requested for his Virginia application.)[71]

Notwithstanding Randel's many supporters, Letitia probably knew she was marrying an irascible man. Perhaps she even relished the controversy. From almost all records that survive—mostly court documents relating to lawsuits or land—Letitia appears to have been Randel's active ally in the quarrels to come. She pushed him to pursue connections. She appears to have been legally and financially astute. She assisted him with business. She read his letters and when needed took action. Writing from Owego, New York, she informed one of Randel's assistants that a lawyer had outstanding questions about the canal lawsuit. She instructed the assistant to travel to New Castle, Delaware, to procure documents and answers. She also provided an update on Randel's work in Owego: "The remaining twenty miles of road to be graded, is to be contracted for to-morrow. This keeps Mr. Randel very much engaged; he will, I hope, after to-morrow, have a little leisure, and write to you, and also attend to Stancliff & Draper's account. Do write, and let us know if you see Mr. Clayton, and what is to be done. Will the suit be tried this August term, or will the cholera prevent the court sitting? Our best respects to your mother and brother, Very Respectfully, your friend."[72]

Letitia oversaw other assistants as well. Her mother, Sarah, came to live with her and Randel in Wilmington, Delaware, at one point and said she often saw Letitia paying Randel's men. Letitia "is very much in the habit of taking receipts," Mrs. Massey reported, but she did not take them from the young gentlemen, as "she thought them honorable, and there was no use for them." Randel relied on Letitia's opinion about employees. "I mentioned to Mrs. Randel that you spoke of employing B. Newcombe; she thinks he is intemperate; I do not know him. If you have not already employed him, you had better employ some other person more eminent at the bar," he wrote to his secretary.[73]

Letitia helped with a clay business that Randel, ever entre-prising, tried to start. He intended to send clay from Delaware to Philadelphia, where brick making and tile manufacturing were thriving. He had a tub built for clay in New Castle but needed iron bands to hold the tub together. He asked an employee to buy some iron and send it down by steamboat, and then left the matter in the hands of his wife: The iron can be sent "with a letter to Mrs. Randel, so that she may have it sent at once to the blacksmith." With Letitia's help, he did send at least one shipment of clay to Philadelphia. There appear to be no records describing how the business fared.[74]

In addition to aiding her husband's business when needed, Letitia ran the household, which had at least one servant, and bore at least three children. John Massey Randel was born on July 14, 1831. When he was two, Letitia and Randel had him vaccinated for smallpox. In the fall of 1833, shortly after that smallpox scare, another son arrived—Richard Varick DeWitt Randel. And a daughter, Letitia Massey Randel, was born on May 24, 1835. Letitia and Randel sent their son John to school in Wilmington and

Title page and poultry diagram from *American Domestic Cookery* by Maria Rundell. Courtesy of the American Antiquarian Society.

Philadelphia, and then to study medicine with his uncle, Walter Williamson, at the Homoeopathic Medical College of Pennsylvania, where John wrote a thesis on hydrophobia, or rabies. Only John survived to adulthood. Richard died in July 1834, at the age of ten months; Letitia died in January 1837, at the age of nineteen months. Randel and Letitia may have had another son, Alexander McLeod Randel (in all likelihood named after a well-known Presbyterian pastor), who would have been six when he died in 1834.

A decade or so after her marriage, Letitia bought or was given a copy of *American Domestic Cookery Formed on Principles of Economy for the use of Private Families*, by an "Experienced Housekeeper" whose non-de-plume name was Maria Rundell. The book was first published in 1822; Letitia dated her copy 1839. None of the pages are folded, none have particular or revealing stains, none imply a well-loved dish. But the book offers a glimpse into what Rundell thought about the women of her time. The words may have resonated for Letitia too. "In the variety of female acquirements, though domestic occupations stand not so high in esteem as they formerly did, yet, when neglected, they produce much human misery. There was a time when ladies knew nothing *beyond* their own family concerns; but in the present day, there are many who know nothing *about* them . . . United with, and perhaps crowning all, the virtues of the female character, is that well directed ductility of mind, which occasionally bends its attention to the smaller objects of life, knowing them to be often scarcely less essential than the greater," Rundell explained in her introduction. She encouraged prompt payment of tradesmen, aptitude with figures, and particular attention to dinner: "Perhaps there are few incidents in which the respectability of a man is more immediately felt, than the style of dinner to which he accidently may bring home a visitor."[75]

If Letitia followed Rundell's advice and recipes, the family had some wonderful meals. Ducks abounded near Letitia and Randel's Maryland home, as John James Audubon described in a visit to the area: "The number of birds set in motion becomes inconceivable, and they approach the points so closely, that even a moderately good shot can procure from fifty to one hundred ducks a day . . . the innumerable ducks, feeding in beds of thousands, or filling the air with their careering, with the great numbers of beautiful white swans nesting near the shores, like banks of driven snow . . ." A few of those ducks may have found their way into a particularly enticing Rundell dish: "Half roast a duck; put it into a stew-pan with a pint of beef-gravy, a few leaves of sage and mint cut small, pepper and salt, and a small bit of onion shred as fine as possible. Simmer a quarter of an hour, and skim clean; then add near a quart of green peas. Cover close, and simmer near half an hour longer. Put in a piece of butter and a little flour, and give it one boil; then serve in one dish."[76]

Shortly after Letitia and Randel's wedding—indeed, that same month—canal commissioners in Pennsylvania asked Randel to direct a survey along a route called the Susquehanna North Branch. Although Letitia often traveled with Randel, just as Matilda had, she did not accompany him on this trip, perhaps because the conditions were rough. The crew walked 365 miles, worked eighteen- to nineteen-hour days, and many became severely ill. On the trip Randel met Charles Ellet, who was then in his teens and had not yet earned his reputation as an excellent engineer. Ellet—whose temperament was, as noted, similar to Randel's—found his boss "a shrewd, calculating, close observing man, and withall a sociable, pleasant, agreeable companion who can convert a word and even a look to answer his own purposes, to be a source of inspiration to other and greater matters than those to which they often relate." By the end of the trip, Ellet championed Randel to his family: "Still in the face of every disadvantage has the principal part of the work been accomplished, by the skill, industry, experience and perseverance of Mr. Randel. And after all the privations and hardships which he has undergone for the people of Wilkes Barre his only thanks have been hard thoughts and his only reward ingratitude. So difficult it is for a man <u>once persecuted</u> to ever find peace again. Our reception here may appear otherwise in the papers, but they do not relate the facts. <u>These</u> are things for the <u>family only</u>. I would not have them mentioned; nor indeed any words of mine where Mr. Randel's name is used."[77]

His Skill, Zeal, Integrity, and Masterly Knowledge

That Pennsylvania survey was Randel's final canal-related job. The Railroad Age was rapidly replacing the relatively brief Canal

Surveyors' camp. Detail from the 1827 "Map of the Canal Route between Pittsburgh and Conneaut Lake." Record group 17, records of the Board of Canal Commissioners. Courtesy of the Pennsylvania State Archives.

Age. Railroads had originated in England in the 1820s and soon traveled across the Atlantic. The first U.S. rail lines were short, local projects, part of an experimental frenzy nearly every state engaged in. Between 1830 and 1840, railroad mileage in the United States grew from 73 to 3,328 miles, and by the decade's end only four of the twenty-six states lacked rail lines. "The most fanatical railroad partisans writing in the 1830s, men like D. K. Minor of the *American Railroad Journal*, could not predict the speed of trains, length of lines, size of locomotives, volume of traffic, revenues, or sums of invested capital that would characterize the industry by the time of the Civil War," writes John L. Larson, the author of *Internal Improvement*. In the early 1860s the transcontinental, first advocated by the businessman Asa Whitney in 1844, began to stretch across the country from California, with the construction of the Central and Union Pacific lines. By then the eastern and central parts of the country had almost 31,000 miles of rail.[78]

Randel, mesmerized by new technologies, mechanical innovation, and a challenge, found employment on several of the country's early local railroads: the New Castle & Frenchtown, the Ithaca & Owego, the Lykens Valley Coal Company, the Central Rail Road of Georgia, and the New-York & Albany. His first appointment came in April 1830. William D. Lewis and Samuel Nevins, two directors for the New Castle & Frenchtown, appointed him chief engineer at an annual salary of $1,500. The company's intention was to compete head-on with the Chesapeake & Delaware Canal and to provide travelers with a faster, more comfortable trip between Philadelphia and Washington.

The New Castle & Frenchtown directors and Randel read a great deal about British railroads. They discussed alternatives and debated during each stage of construction. Randel's survey seems to have been the only aspect of the project not scrutinized too intensely. Randel ensured a largely level path (along an old turnpike) and a route of gentle curves, "enabling us to pass through them without abating [the train's] velocity." He had his men set down 58,000 sleepers—bases to set the tracks on, similar to ties—of stone and later wood, because stone took so long to obtain. The crew spiked iron rails from England on top of wider, supporting rails of yellow pine. Randel used what became the standard gauge.[79]

Work progressed quickly. Within two years of Randel's appointment, horses pulled the first carriages along the rails. The February 1832 trip between New Castle and Frenchtown took one hour and twenty minutes; about half as long as it would have taken by

Stock certificate from the New Castle & Frenchtown Railroad. Courtesy of the collection of Mark D. Tomasko.

stagecoach and one-third as long as it would take a wagon loaded with goods. The board of directors then instructed Randel to rapidly ready the line for the locomotive, which would soon arrive from England. (It was assembled with difficulty by Philadelphia engineer and jeweler Matthias W. Baldwin, who went on to design and construct many early American railroads. Studying British designs and imports was instrumental to his success.) To ready the track for the demands of a steam engine, Randel hired more men; at one time, 1,100 were working night and day. He reported to the directors that he too worked round the clock and "gave the whole my personal attention for upwards of 20 out of the 24 hours of each day."[80]

The first locomotive successfully ran just a few months later, on July 4, at an average of 12 miles per hour. Occasionally the train reached speeds of 30 to 40 miles per hour. For Randel, achieving such velocity on the first run portended greater speed to come: "I have no doubt that the whole distance of 16 miles and a half from New Castle to Frenchtown, can be passed over with this Engine and tender, in the short space of 20 minutes, or at the extraordinary rate of 50 miles per hour; a speed far surpassing, and perhaps trebling the velocity which for some time to come will be agreeable to the passengers crossing this peninsula."[81]

Randel also submitted his final report on July 4. As always, he had several ideas for improvements. He suggested installing a "switch with its excentric cam or wheel (upon a plan believed to be entirely new) to be put down at the crossing of the rail way near, & west of the Engine house at New Castle"; he noted that he had already given a mechanic his plan for the switch. (Such switches enable trains to move from one track to another.) He wanted safety wheels installed as backup in case car wheels broke: "I have caused an example exhibiting the *principal* of this Improvement to be applied to one of the Cars for your examination." He had also nearly completed a "detaching link or lever," so the crew could release the passenger cars from the engine if needed (in case of accident, or perhaps boiler explosion, which was common in early engines). He "had prepared an example" of fenders and cushions to be fixed to the end of the carriages to soften the ride. In addition, "a Scraper to be attached to the front of the Engine or forward Car for the purpose of removing all obstructions from the Road (even Cattle) has been invented and planned, but want of time will deprive me of the pleasure of completing it at present." His list of improvements concluded with a generous offer: "If the Directors request it, I will give them my permission to use the above improvements upon this Road."[82]

Contemporary and modern views vary on Randel's accomplishments on the New Castle & Frenchtown, Delaware's first railroad. The most positive assessments came from Randel himself, his great-nephew, some news accounts, and the directors of the company. A negative evaluation comes from historian Larry D. Lankton, who reviewed the railroad for the National Park Service in 1976. According to Lankton's report, Randel and his cohorts had little idea what they were doing; Randel managed his workers ineffectively and made poor decisions; the entire enterprise was unprofessional and haphazard. The directors were heavily involved with the daily activities; they recognized their own lack of knowledge and needed to be thoughtful at every costly step—and they were aware of Randel's lack of engineering experience. "His subsequent, mediocre performance only served to demonstrate that surveying and engineering were two related but different skills," Lankton writes.[83]

Lankton's assessment echoes descriptions of efforts on other inaugural U.S. railroads. Randel was not alone in his inexperience. No one much knew what to do, and the New Castle & Frenchtown was among the first rail lines to experiment with the form and, particularly, with locomotive as opposed to horse power—that is, having horses pull cars along the rails. Just as they had on the Erie and on many other canals, pioneer U.S. railroad engineers bricolaged, importing some techniques from England, making others up. Various companies and their engineers experimented with track designs and materials, with gauges (the distance between the two rails, which today is standardized in most places at 4 feet, 8.5 inches, but which varied widely until the late nineteenth century), with horse-drawn cars, and with British and U.S. engines.[84]

George Johnston, who wrote a history of Cecil County, Maryland, in 1881, described the New Castle & Frenchtown railroad as an oddity: "It was of very peculiar construction, and were it now extant, would be a great curiosity. The rails were placed about the same distance apart as in modern roads, but instead of being laid upon wooden sleepers, were placed upon blocks of stone ten or twelve inches square . . . The great defect in the road was the want of something to keep the rails from spreading apart, and it was soon discovered that the only way to remedy this was to resort to the use of ties extending from one rail to the other, and to which both rails were fastened, as in modern roads."[85]

For other historians, the railroad was progressive for its early use of a steam locomotive and for its use of semaphores. John C. Hayman credits the New Castle & Frenchtown with pioneering flag signals: "As the train left a terminal a white flag was hoisted, a black flag if it was late or became disabled." A train worker down the line would see the flag in his telescope and pass the information to the next point, "so that news of the train's departure reached the other end in very short time."[86]

In his own day, Randel earned praise from the press and his employers. "Since the commencement of operations upon our Rail Road, those who have traveled upon it, so far as our information extends, have spoken in very favorable terms of the plan and execution of the work. The Board of Directors, considering that both are the result of your skill and labour, desire that you will receive this letter as a testimony of their high estimation of your talents, ability, and skill as a Civil Engineer; of their thanks for your industry, zeal and fidelity in prosecuting the work; and of their respect for your character and correct deportment as a man." The timing of this public support for Randel was notable. The Chesapeake & Delaware Canal Company lawsuit was approaching the courtroom of Judge Harrington, and Randel's "industry, zeal and fidelity" would soon be under intense legal scrutiny.[87]

The directors' support of Randel did not waver as time went on, although the New Castle & Frenchtown itself was financially shaky by the 1840s because of competition from the Philadelphia, Wilmington & Baltimore Railroad. When Randel sought employment on the Harlem Rail Road in New York City in 1845, his former employers endorsed him: "His skill, zeal, integrity, and masterly knowledge of his profession, carried him through that important enterprise at so early a period in the construction of Rail Roads, in a manner to satisfy us entirely, and to place that work, even at the present day, in advantageous contrast with almost all existing Roads of the kind," wrote William D. Lewis and Samuel Nevins.[88]

Fences Are Prostrated

Other contemporaries also perceived Randel as able—or as able as anyone else at that time. In the midst of his work for the New Castle & Frenchtown, he was hired by another early line, New York's Ithaca & Owego, for $3,000 a year. Simeon DeWitt and his son Richard Varick DeWitt held the most shares of stock in the Ithaca & Owego and naturally turned to Randel. Ithaca was connected to the Erie Canal, which ran north of the city, by Cayuga Lake and the Seneca River, but it was not linked to several important markets and suppliers—such as Maryland and Pennsylvania—except by a road running south to Owego, a town on the banks of the Susquehanna River. Both Owego and Ithaca wanted to solidify their roles as important trade centers mediating cargo coming out of the west and east. As an Albany newspaper hopefully summarized, "Ithaca in fact is the key to the trade of the upper counties of the Susquehanna, and distributes salt, plaster, castings, and merchandise, to a great section of country in Pennsylvania. It receives lumber (the finest that comes to this market), produce and coal in large quantities and will furnish an inexhaustible supply of fuel for the furnaces and salt works of our state."[89]

Initially town leaders proposed a canal, but they could not secure funding. In 1827 citizens called for a railroad, which was incorporated a year later; a U.S. Army engineer was immediately brought in to survey a route. The DeWitts and the board then hired Randel as chief engineer. "Mr. Randel, Chief Engineer of the Newcastle and Frenchtown Rail Road, has commenced active operations on the Ithaca and Owego Rail Road," the *Albany Daily Advertiser* reported in July 1831. "The great facilities for making this road, and the admirable nature of its route, surpass all previous calculations. A small part of the road will be made this season and contracts made for the residue this fall, to be completed during the next season. Real estate is rapidly advancing at the points of termination, and capitalists are already seeking investments in the vicinity."[90]

Construction of the New Castle & Frenchtown and the Ithaca & Owego overlapped, and so Randel raced back and forth between upstate New York and Delaware. He had two unpleasant experiences during this time. In early 1832, a few weeks before the New Castle & Frenchtown line was tested, he was in a stagecoach accident. He injured his head, arm, and hand. In July 1833 a violent storm, with 4-inch hailstones and "sixteen of them weighing about a half a pound," brought danger and death. "This fine country, which a few hours since exhibited a most beautiful appearance, is now completely destitute; fences are prostrated; many trees torn up, and some buildings blown down and destroyed," reported the *New-York Spectator*. The tavern-keeper Albert Johnson was killed, and Randel nearly so. "The deceased, together with J. Dandel [*sic*], Esq. the Engineer in Chief of the Ithaca and Owego Rail-road, and Mr. Tollfree, one of his assistants, were standing in conversation near the house when the storm commenced, and returned for shelter to the carriage house, both doors of which were open. The door of this building was struck by lightning, and all three of the persons prostrated. Mr. Johnson was instantly killed. Messrs. Randel and Tollfree so far recovered as to be able to rise in about one minute from the time of receiving the shock. The latter was slightly, the former was seriously injured, but has now nearly recovered from the effects of the shock." An *Albany Daily Advertiser* account noted that a horse standing alongside the three men was killed as well, and that Randel and his assistant had been accompanied by the president of the rail line, Francis A. Bloodgood, who ran into a shop instead of the carriage house and suffered no ill effects.[91]

True to his nature, Randel offered the directors five possible design plans and cost estimates. In the end, the company chose horsepower over steam, so that the line would be cheaper to build. Randel's work took him through valleys and gorges, across many creeks, around falls, and up and down the renowned hills of the area. He had to use a deep cut through one hill and construct several culverts and bridges. The terrain was challenging and quite unlike the level, smooth land through which the New Castle & Frenchtown traveled, and the line was relatively short—only about 29 miles. With seven hundred men on the crew, work was completed within two years. Labor often proceeded into the night, reported the *Ithaca Chronicle,* "and the woods have been illuminated by hundreds of candles sparkling along the line." The first half opened in February 1834; the rest was finished two months later. Ithaca was proud, reported the *Chronicle*: "Much of the road is finished in a durable and beautiful manner furnishing perhaps the finest specimen of railroad in the union."[92]

Although Randel sought, and said he obtained, "the *most eligible routes*, *grades* and *curves* the most *gentle,* with straight lines connecting them of the *greatest length* that the country would afford," there was one elevation he could not tame. The outset of the route at Ithaca was anything but eligible: a stunningly steep hill, more than 500 feet high. This ascent, South Hill, was the first hurdle that horses, and later engines, had to clear. Randel installed a system of horse-turned winches to haul up the freight cars. "The horses went round and round like those that work a threshing machine," described a contemporary who worked on the railway. "The cars were let down and hauled up the high, steep hill by that windlass-like system. While two cars were down it aided in hauling one car up the plane. A man went along with them carrying oak plugs to use as brakes in case the rope cable broke. The plugs were thrown into the car wheel spokes and caught the wheel against the car."[93]

Horses also labored at the Owego terminus. There, the same rail worker described a wild scene: "Four horses worked this windlass down in a pit . . . the belly-bands of the harness were wide and strong and often held the horse clear up from the floor when the cars got under too rapid headway on the steep plane and held them suspended in that position until the cars reached the level and ran into the car houses and were stopped by men who threw oak plugs into the wheels. Then the horses were lowered again to their feet."[94]

The steep hills were often miserable for passengers as well. The descent was dangerous, and on at least one occasion the handbrake and oak plugs failed. Passengers had to leap from a car as it careened "like a cannon ball" to splinter at the base of the hill. No one died in that accident, including the one sleeping traveler, a Mr. Babcock, who "eventually recovered" from his broken arm, many bruises and cuts; but several engineers died in other accidents. Passengers were routinely called upon not just to save themselves but to help the line run smoothly. Initially the line had no signals, no schedule, and when trains heading in opposite directions met, one of them had to be lifted off the tracks so the other could get by. The passengers did the lifting.[95]

All early railroads were notoriously unsafe. The metal strap rails that Randel used could spring off the wooden rails they were attached to. These so-called snake-heads could pierce the bottom of the cars, impaling passengers. Cars jumped off rails, boilers exploded, trains hit pedestrians. Randel's own family suffered.

The Inclined Plane of the Ithaca & Owego Railroad. Painting by W. Glenn Norris. Image courtesy of the History Center in Tompkins County, Ithaca, New York.

When a rail line was laid between Utica and Syracuse (the line later became New York Central), Abraham, Randel's older brother, received $240 for the right of way across his land and a lifetime pass for himself and Rebecca. But the proximity of the track proved fatal. On the morning of November 7, 1856, Abraham was hit by a freight train; he died that evening, at the age of seventy. The New York Central also killed his granddaughter. "One of the saddest railroad accidents that has happened in Oneida in many years and one that will be fresh in the minds of the people for some time to come, occurred last Saturday evening," reported a local newspaper in 1895. Helen E. Randel, a "popular young school teacher," was suffering toothache and set off for town in search of a dentist, accompanied by her fiancé. The family often used the tracks, which ran right by the property, to walk into town or go to church. Passenger train 47 hit Helen and her fiancé, killing both.[96]

RANDEL SAW the Ithaca & Owego as "destined to become one of the most important links in the chain of internal improvement that has yet been projected in this section of country." But like many early lines, the railway did not have a long life. The horsepower-based design meant that travel along the rails remained quite slow—about walking speed. The line finally incorporated engines in 1840; Richard Varick DeWitt designed the first one. ("Old Puff," as the engine was called, was soon replaced by a more effective machine.) According to one historian, the Ithaca & Owego line was an important proving ground for U.S. railroads, as much for its failures as for its successes. The tracks, it turned out, could not well bear the weight of steam engines, and repairs were constantly necessary. The cost of the repairs led the company into foreclosure in 1841. The line was bought and incorporated, over many decades, into a series of other railroad companies. The Ithaca branch, as it came to be called, served the region until 1956. Clarity about the best approaches to railroad construction and design came slowly through experiments, like Randel's, and with hindsight.[97]

Randel's employment on two of the country's first railroads solidified his reputation as an engineer. Barely was he done with the Ithaca & Owego when a Georgian concern requested his services. The board of directors of the Central Rail Road and Banking Company sought a route from Savannah to Macon and "a gentleman of high standing and great experience in his profession" as engineer. The gentleman of high standing and great experience

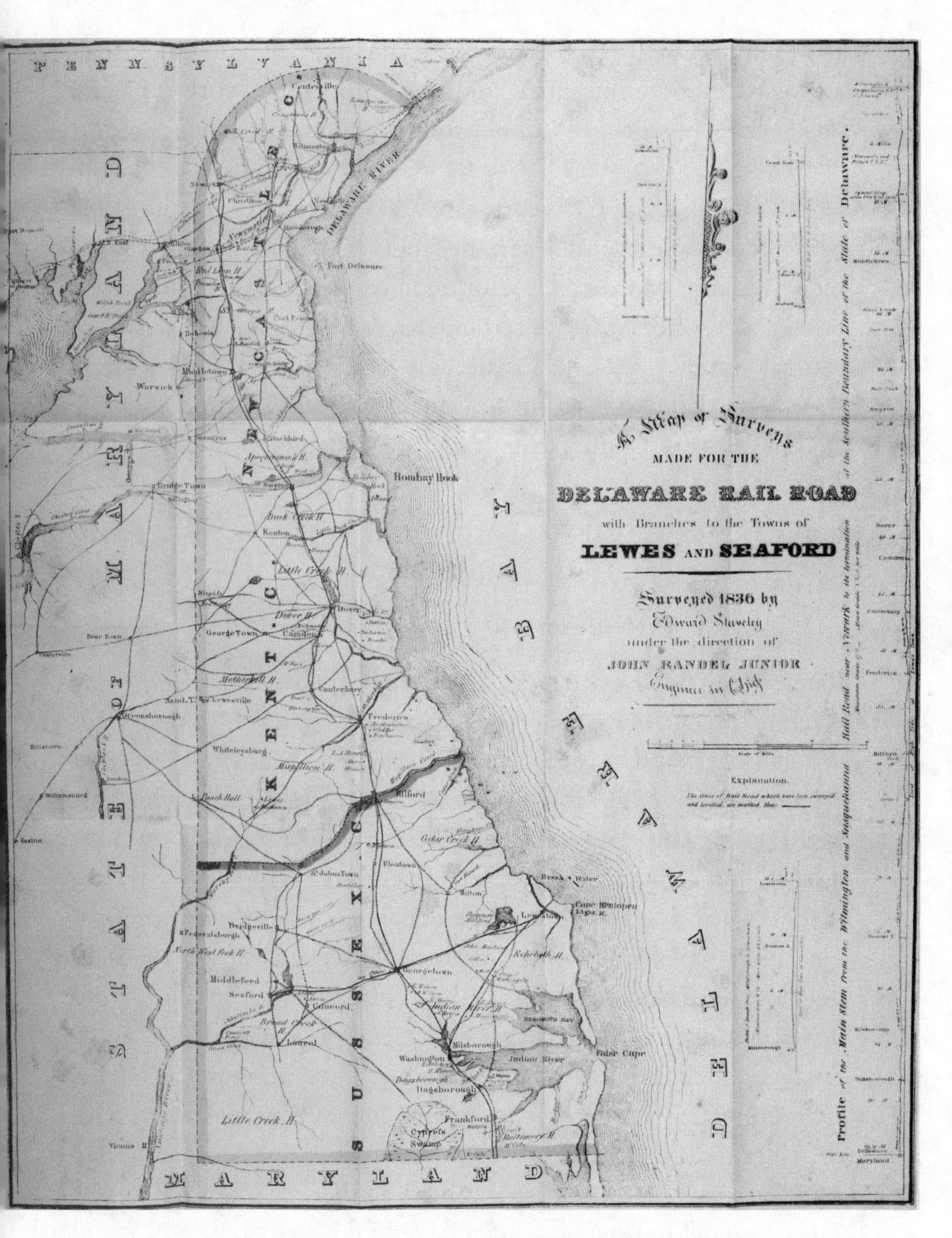

Map of the proposed Delaware Railroad by Edward Staveley and John Randel Jr. Courtesy of the American Antiquarian Society.

soon differed with the gentlemen of the board. Randel, as usual, advocated a route that made eminent sense from a surveying and engineering perspective but not from a political or economic one. "Perhaps the key to reasons for Randall's dismissal in May (in spite of the fact that for engineering reasons the southern route was palpably the best route) lies in the fact that, as his successor Reynolds says, the board of directors, for reasons of a non-technical but equally valid nature, entertained a preference for the northern route," summarized one researcher. Randel's work in Savannah was brief, lasting only from November 1836 to late spring 1837. And bitter—it was in Savannah that his daughter Letitia died.[98]

I Am a Ruined Man

Randel learned a great deal about railways during those three adventures, and during short stints as surveyor for the Lykens Valley and Coal Company (a Pennsylvania line), as engineer in chief for a proposed Delaware Rail Road Company (founded by his lawyer, John M. Clayton), and as engineer in chief for the New-York and Albany Railroad Company. He seemed entranced by trains; several years later, his grandest mechanical scheme grew directly out of his earlier railroad work. But as pleased and proud as Randel must have been to know that his reputation for accuracy, if not diplomacy, was intact and that he was still well regarded by some employers, he doubtless felt encumbered and anxious during his railroad jobs. There was no escaping ongoing entanglement with the Chesapeake & Delaware Canal Company. The settlement had arrived in early 1834 but had brought neither payment nor peace. The company refused to—indeed, could not—pay him.

Encouraged by his lawyer, Clayton, and supported by the State of Delaware, Randel took matters into his own hands. In June 1834, he announced that he would collect canal tolls himself—"attachment," in legal language. Though he resisted the idea at first, Randel came to eloquently and energetically embrace the strategy. The company had forced his hand. Days after the case concluded, the board of directors resolved to require all ship captains to pay the canal toll in Philadelphia. They did not publicize their resolution—probably because it was illegal, according to the charter governing the canal, and because the Delaware court had ruled that Randel's payment was to come, in part, from tolls. And so Randel's agents began demanding tolls on the canal itself, much to the dismay and surprise of boat captains who had already paid them.

On June 15, Richard Shoemaker had the misfortune of being arrested by the Newcastle County sheriff at the Delaware tide lock, the location established in 1829 for toll collection. Shoemaker was the captain of the sloop *Robert and James*, and he had already paid $74.44 to an officer of the canal company in Philadelphia. Shortly thereafter, Thomas P. Crowell, master of the schooner *Hiram*, fared no better. He had paid $96.28 in Philadelphia. Caught between the company and the law, captains found themselves having to pay twice—and being "artfully induced to believe that I was their oppressor," Randel wrote. Shoemaker and Crowell were among dozens of captains similarly treated. One witness estimated that attachments were served on as many as 1,500 people.[99]

Randel's piracy was administered by a young man named William Linn Brown. In the early days of the Chesapeake & Delaware lawsuit, the twenty-five-year-old Brown had arrived to study law in the office of Philadelphia lawyer John Sergeant—the one whom Carey had hoped to influence regarding strategy. Sergeant represented Randel at the beginning of his troubles with the company, and so Brown got to know Randel and often stayed with him. "Mr. Randel told me that Mr. Brown was his personal friend, the nephew of Mr. De Witt [*sic*], and that Mr. De Witt, Surveyor General of the State of New York, was Mr. Randel's patron, and therefore he had the utmost confidence in Mr. Brown," said an observer. When Randel switched from Sergeant to Clayton as his main counsel, Brown stayed with him. Brown was apparently promised 2.5 percent of whatever settlement as well as reimbursement for all his traveling expenses—of which there were many, as he darted around the region at the behest of Randel or Letitia. Brown devoted himself to Randel "against the advice of all his friends" except Simeon DeWitt; "Mr. DeWitt thought Mr. Randel's cause was a good one, and that he would gain it," according to later testimony. Several people described Brown during that time as "the second edition of Mr. Randel . . . He manifested a very unusual interest in the affairs of Mr. Randel. I never knew a man take such interest in another man's business. He employed himself actively at his business. I mean with zeal."[100]

Brown spent a great deal of time on the canal. According to a deputy sheriff, "Mr. Brown was engaged there; he was with us assisting in issuing attachments," basically writing tickets to boat captains.

> He was there, I suppose, a month at a time . . . We were pretty constantly employed; sometimes night and day; sometimes we had nothing to do; but we had to be always there, that is, while he was issuing the attachments . . . at all hours of the night, if we wanted a *capias* issued, we had to go to Mr. Brown's room to have it issued . . . We had a watchman. Whenever he saw a vessel coming, he would wake me up. I then went to see if the captain was liable; if he was liable, I called on Mr. Brown. The most that I ever called was four times in one night . . . We had pretty nearly all the country about there to contend against; and to stop the captains, we used to set up till about twelve o'clock at night . . . The country was very sickly at that time—all hands got sick—Mr. Brown got sick—all the deputy sheriffs became sick—myself too . . . the night air was dangerous, and the smell was bad—filth from the stables.

The sheriff jailed many of the captains who did not pay Brown.[101]

Although Brown's nocturnal activities were supported by some authorities in Delaware, the strategy was controversial. Through captains Shoemaker and Crowell, the canal company brought two more lawsuits against Randel. For their part, Randel, Brown, and their team tried to establish further legal support for their actions by promoting bills permitting attachment, based on laws in Massachusetts and New Hampshire. The bill passed the Delaware senate but was defeated in the house. "I am a ruined man," Randel wrote in a local paper, "but however crushed and powerless I may be, I yet have a right to appeal, and will fully appeal, to the justice of the public." Randel was again airing his grievances and taking his problems to the people, as he had with William Bridges and the commissioners of the Erie Canal. He begged forbearance on the part of his creditors and contractors, and concluded his article by saying, "I have come out of this controversy, pennyless and exhausted by it, but yet with a breast untainted, and a heart unstained."[102]

Soon thereafter, the Delaware Superior Court found for Randel in the cases brought by Shoemaker and Crowell. Still no peace. The following month a court officer delivered to Letitia, at home in New Castle while Randel was working in Syracuse, a notice from the U.S. Supreme Court requiring Randel to appear on the second Monday of January 1836. Randel returned from New York and went to court, again represented by Clayton. The canal company had sought to overturn the Delaware court's decision regarding the captains, and it had sought to overturn the original ruling in which Randel was awarded the settlement. The U.S. Supreme Court was having none of either. First it rejected the company's effort to overturn the 1834 case. Then it rejected the company's arguments to overturn the attachment cases. "This decision leaves the canal company no resource but to pay the money which has been awarded to Mr. Randel by the verdict of a jury," reported the *Wilmington Journal*. "The highest judicial tribunal of this state has established the right of Mr. Randel to appropriate the tolls for the payment of his debt, and the supreme court of this union now tells them that they can obtain no relief there." The editors went on to state, "We repeat, then, as the best advice we can give to the company—and we give it in a perfect friendly spirit—pay this money—put an end to this strife which has for so long a time embarrassed the business of the canal and destroyed its usefulness, and which if persevered in, can only end in the total ruin of the interests of the stockholders and creditors."[103]

A few months later, fed up with the canal company, the Maryland and Delaware state governments passed bills requiring the company to settle its outstanding debts within five years. Randel had been vindicated. But he was up to his ears in debt. Nothing had come of the Georgia rail work. He had done a topographical and hydrological survey for the city of Baltimore, which was seeking a more reliable source of municipal water, but he was not hired to build the system; the city council decided that the cost was prohibitive. As one friend noted, "I believe the canal debt is his sole reliance. I have been intimate with Mr. Randel for some seven or eight years, and he has spoken very freely to me of his wants. I have tried to borrow money for him on pledge of certificates, and on mortgage of his real property in Delaware and Maryland, but ineffectually." At the same time, friends and colleagues were increasingly questioning his character and deportment. Letters from this era and volumes of court testimony portray Randel as sometimes paranoid, volatile, and unpredictable. He was sued for payment by several former allies. Perhaps the most tragic reversal involved his relationship with the young man Brown. Long-standing loyalty and friendship turned to bitterness and betrayal. And another series of lawsuits began their voyage to the U.S. Supreme Court.[104]

It Is Equally Uncertain Whether My Destiny Lies the One Way or the Other

Succinctly put, Randel accused Brown of absconding with $10,000 and a power-of-attorney document. Brown countered that he had not been rewarded for his many years of service, and that he had not absconded with anything Randel had not legitimately given him. Court documents from two combined cases (*Randel vs. Brown* and *Brown vs. Randel*) include testimony from seemingly every person who met or dealt with either man between 1825 and 1839. Letitia's mother testified. Randel's friend from New York City testified. Sheriffs, lawyers, captains, workmen, and acquaintances testified, sometimes dramatically: "The said complainant, with much sharpness and temper, said, if he would not sign such a receipt, he would compel him. And getting still more excited, he struck his clenched hands violently upon the table and exclaimed, '*I need not have paid you one dollar.*'" Randel lost in a Pennsylvania court in 1841 but won on appeal to the U.S. Supreme Court in 1844. Brown, who had devoted at least a decade of his life to Randel, had been turned out and rejected. Randel, whether in the right, unable to admit wrong, or miserly, gained nothing in the eyes of his acquaintances. Richard Varick DeWitt wrote to John M. Clayton expressing wariness about his own financial agreements with Randel and requesting advice. "Please to regard this Communication as Confidential," he concluded. "You understand Mr. Randel's peculiarities too well to make it necessary to say why I ask this."[105]

Richard Varick DeWitt's father was spared whatever sadness the showdown between his nephew and his protégé might have elicited. In December 1834, just shy of his seventy-eighth birthday, Simeon DeWitt died after a long illness, "a violent cold" he had caught when traveling upstate. Randel had visited him in October, which may have been the last time the two men saw each other. DeWitt was weak, able to walk only across his room. "I still stand poised between Life and Death, under my present disease, and it is equally uncertain whether my destiny lies the one way or the other. Whatever it may be, I am perfectly reconciled to it," he wrote.[106]

In his seventy-seven years, DeWitt had observed and participated in the radical transformation of his country, culture, and landscape. His land had become American. His state had prospered, in large part because he facilitated settlement with his careful grids and maps and because he fought for and funded infrastructure. He had established important cultural and scientific institutions in Albany and participated in those of other cities. He had given a powerful and interesting city "harmonious" proportions. He had trained many young men to be topnotch surveyors and scientists. He had given generously of his friendship and good humor.

Although fond and supportive references to Randel had continued to appear in DeWitt's writings, some tension had entered the relationship in the 1820s and the early 1830s. Field books are official records, but Randel saw right to purloin some upstate New York field books—to treat them as his private property, complained state officials. Controversy regarding the boundaries of lots in the Onondaga Salt Springs Reservation arose as early as 1824, and DeWitt entreated Randel to send the books back so boundaries could be clarified. "I have never been placed in so disagreeable a situation before," DeWitt wrote. "I must, therefore, beg of you not to delay a moment longer, what is necessary to enable me to have the lots staked off, and to give such descriptions as that patents may be given to those who want them." By 1831 the matter had not been resolved. DeWitt wrote to Randel that complaints continued: "You must not let this winter pass by, without finishing this business, and I wish you to say to me in a letter, that it will certainly be done, in order to satisfy those who may otherwise take measures that might be unpleasant to me. I hope you will let me hear from you soon, about this business." DeWitt would have been displeased, although perhaps unsurprised, to learn that the problem persisted well after his death.[107]

Between 1838 and 1839 surveyor general Orville L. Holley sent many pleading letters to Randel. Initially he intended to hire Randel to do more surveys in the Onondaga Salt Spring Reservation; then he begged for the data already in the notebooks. "Now, my dear sir, let me urge you in the most earnest and at the same time in the most friendly manner to write either to Dr. Green, or to me, or to both, and to forward the notes so much wanted without delay," Holley wrote to Randel in June 1838, ever careful in this and all his other letters to be reassuring and collegial. At that time the New York State legislature required Holley to issue a map and description of all the lots in the reservation being used for the production of salt and those "vacant"; the new survey would necessarily rely on earlier survey notes, which Randel had made many years before. Randel managed to evade Holley's requests for a long time. He said he had been out of town, caught in a lawsuit, and that his wife was ill, which is why he had not responded sooner. Eventually he did release some notes and go to advise in the field. In July, Holley wrote an optimistic letter to Green: "I am glad to hear that the work is drawing so near to a close. Mr. Randel is now with you, I suppose, as I saw him for a moment in the

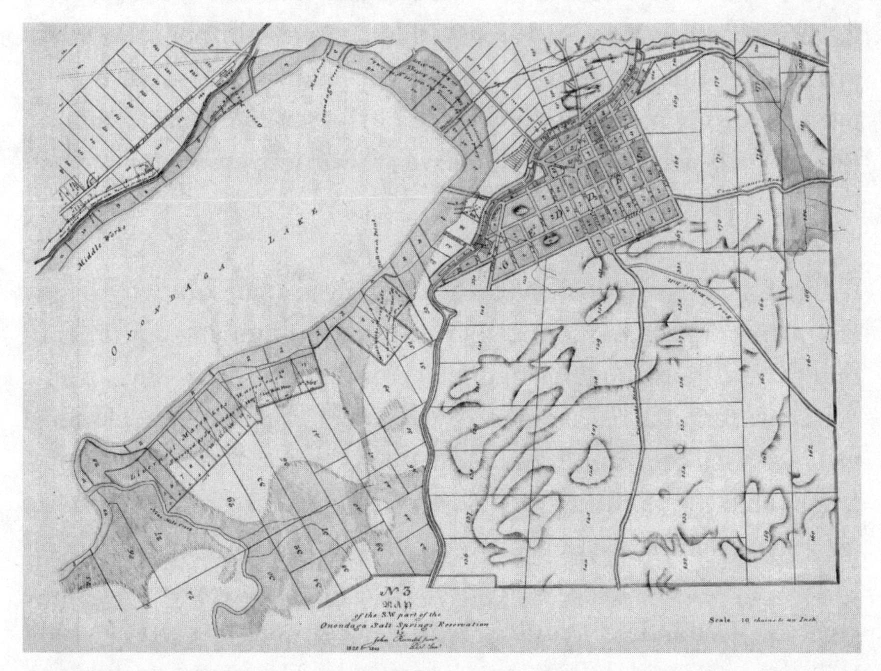

Onondaga Salt Springs Reservation by John Randel Jr. Courtesy of the Bureau of Land Management, New York State Office of General Services in Albany.

street here, the other day, as he was hastening to the Rail Road Station to take passage for Utica, on his way to Syracuse. He has with him in his trunk, as I understood him, everything necessary to enable you to complete the map and descriptions."[108]

It was wishful thinking. By November, Holley knew that Randel retained still other field books. He exhorted Green: "Don't let him rest till he sends them on—press him hard." No luck. Finally Holley sent an assistant surveyor, William F. Weeks, to Maryland to get the notebooks, writing, "You have done wisely in remaining to get such notes &c. as are needed, from Mr. Randel's papers, maps, &c. All I can say is, *for God sake get all* that are needed as speedily as possible, and let us my dear fellows have this Salt Springs Reservation off our hands as soon as the fates will permit." Weeks prevailed, and on February 2, 1839, Holley wrote to Green, "Light shines at last."[109]

Why Randel refused to help or to return documents belonging to the state is not known. He clearly felt that some of his notes were in part private. It was also likely that he hoped he could profit from his knowledge; perhaps he could earn some money because of his intimate knowledge of the land and possession of the notes. Indeed, his resistance was rewarded, for Holley did pay him to assist Green in Syracuse.

RANDEL BECAME a wealthy man when the canal company began paying him his settlement. And yet from that time on, court documents reveal nothing but financial trouble. After the Chesapeake & Delaware Canal case and its myriad related cases, Randel seems to have been constantly in court. Records from New York, Maryland, Delaware, and the U.S. Supreme Court attest to a rich legal legacy for Randel, one usually entailing land, mortgage, and debt. In one New York case, Randel and his nephew, Jesse F. Randel (the son of Randel's brother Daniel), were jointly sued by a Mr. William A. Beecher. Jesse had borrowed $275 from Beecher and Randel had been the guarantor.[110]

Although the canal suit had been atypical in both its duration and its award, Randel's later debt-related litigiousness was not unusual for that time. "Debt was an inescapable fact of life in early America," writes Bruce H. Mann in *Republic of Debtors*, noting that debt cases were the most common form of case in the late eighteenth and early nineteenth century. Attitudes toward debt cases changed over the course of Randel's life—in a way that would have eased his Presbyterian conscience, lifted him off the moral hook. Several legal historians, including Mann, at Harvard University, and Alfred S. Konefsky, at the State University of New York at Buffalo, describe debt cases before and just after the Revolutionary War as largely local, unfolding within a network of interdependent people often well known to one another and governed by moral understanding. During the nineteenth century, markets expanded, cities grew, and familiarity with cycles of boom and bust emerged; Americans lived through the panics of 1819, 1837, 1857, and so on. Debt was no longer seen as the province of the individual and his moral fabric; it became linked to the larger economy and to forces beyond a person's ethical constitution. "The key psychological change for me is the recognition that there might not be moral stigma associated with failure to pay," Konefsky says. "If you believed in the free market, you needed to recognize that there were events out of the individual's control; so they might be overtaken by events."[111]

Many were thus afflicted. "In no country in the world are private fortunes more precarious than in the United States," wrote Alexis de Tocqueville. "It is not uncommon for the same man in the course of his life to rise and sink again through all the grades that lead from opulence to poverty." Debt was a hallmark of many inventors, who sank money into their visions and devices and were often destitute or overdrawn; Charles Goodyear, Edwin Drake, Elisha Otis, and Nikola Tesla are but four examples. The game may be well played, but it is long and unpredictable.[112]

Strange and Eccentric, Full of Utopian Schemes and Projects

In 1858 Simon J. Martenet, a surveyor from Baltimore, published a map of Cecil County, Maryland. The map shows the Chesapeake & Delaware Canal cutting in from the east, its westernmost extent emptying into Back Creek near Chesapeake City, which the canal split in two. Back Creek flows to the Elk River, a tributary of Chesapeake Bay. The area framed by Back Creek to the north and Elk River to the west is called Bohemia Manor. A few years before Randel was born, Simeon DeWitt surveyed the region on George Washington's instruction: "Immediately upon receipt of this you will begin to Survey the road (if it has not been done already) to Princeton, thence (through Maiden head) to Trenton. Thence to Philadelphia, thence to the head of Elk through Darby, Chester, Wilmington Christiana bridge. At the head of Elk you will receive further orders. I need not observe to you the necessity of noting Towns, Villages and remarkable Houses & places but I must desire that you will give me the rough traces of your Survey as you proceed on as I have reasons for desiring to know this as soon as possible." Some of the houses and roads DeWitt noted may have survived to be noted by Martenet. Many were new, constructed as the canal generated a bustling economy. A cluster of those houses, Martenet indicated, belonged to "J. Randall": two buildings on the north side of Back Creek, including Welch Point, and four buildings and a sawmill on the south side of Back Creek.[113]

The canal company earnings had elevated Randel to landed gentry. He bought a farm from James A. Bayard of Wilmington (a lawyer for the canal company) for $12,000 in 1836. In 1837, land records show that Randel owned 740 acres. In 1842 he owned two farms, a steam mill, and about 1,415 acres. He named his land Randelia. Some accounts, including that of a Randel descendent, maintain that he purchased holdings on both sides of the

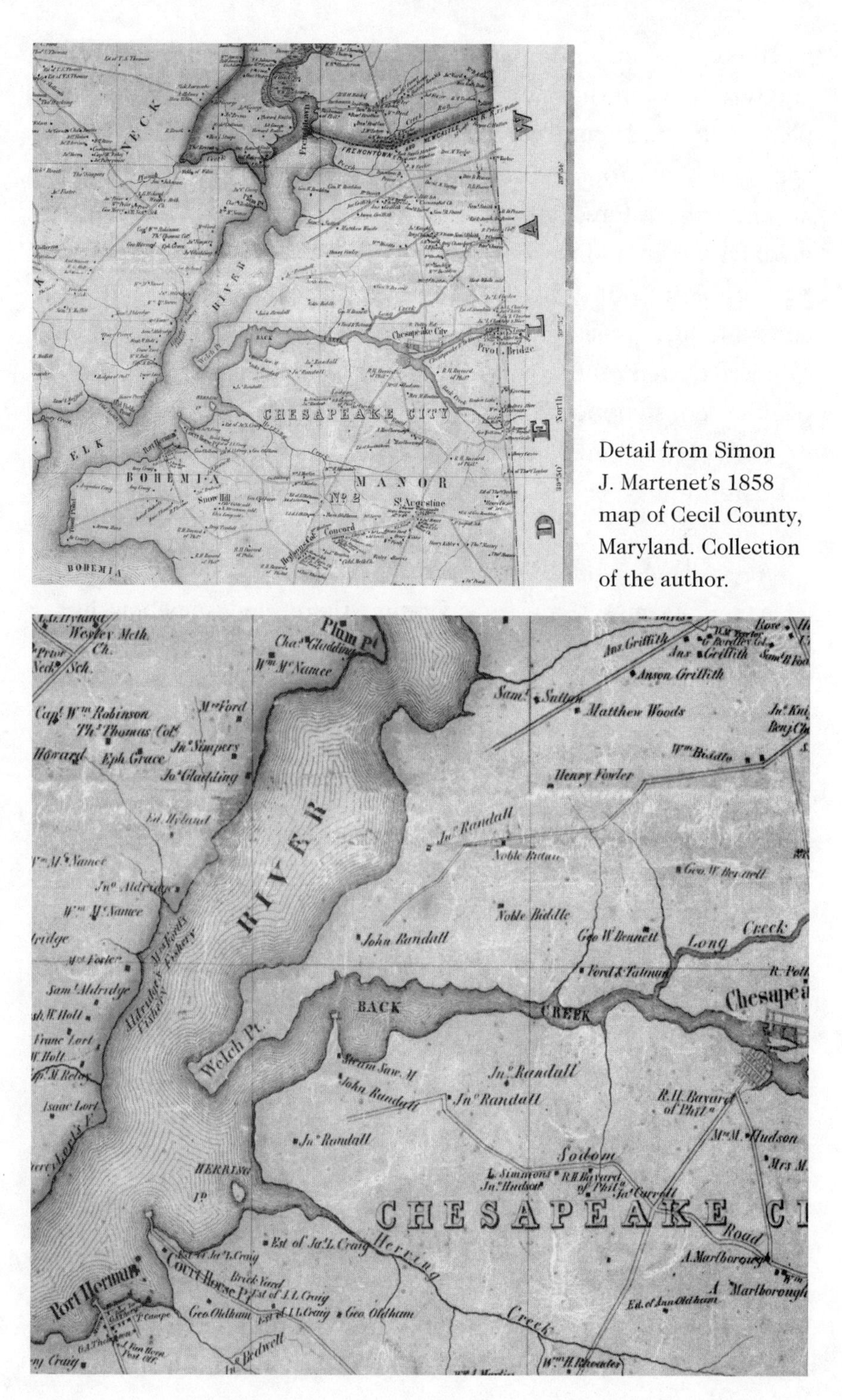

Detail from Simon J. Martenet's 1858 map of Cecil County, Maryland. Collection of the author.

river because he anticipated that the canal and railroad would catalyze a boom, perhaps grow a city to rival New York. Chesapeake City did expand somewhat, but not much. The region never spawned a metropolis, and although Randel sought to establish an enduring estate, only one or two references to Randelia survive. One is in a Maryland land record, and one is in an account by the Cecil County historian George Johnston—not a flattering account, but consistent with Johnston's assessment of Randel's New Castle & Frenchtown railroad as a "great curiosity" and "of peculiar construction."

Randel tried his hand at several businesses, according to Johnston:

> At one time, while Mr. Randel was proprietor of Randalia, he had a steam saw-mill in operation there, and somehow he unfortunately lost a breast-pin which he valued very highly. Work was immediately stopped at Randalia, and everybody in his employ was set to work hunting for the lost breast-pin. The hands at the saw-mill were set to work sifting an immense pile of saw dust, the accumulation of years, on order to find the lost jewel. After much tribulation, the long-lost and much-esteemed bauble was found in the possession of some person, who said he had found it along the road some distance from Randalia, where no doubt his owner had dropped it. The chances for a law suit were not to be lost, however, and the contentious Randel laid his case before the grand jury with the intention of having the person who found the breast-pin indicted for theft, but the grand jury wisely dismissed the case.[114]

According to Johnston, Randel was "strange and eccentric, full of Utopian schemes and projects." One of those projects might lie behind a baffling notice that ran in the *Cecil Democrat* in 1864, the year before Randel's death. The seventy-seven-year-old engineer advertised for three hundred carts and wagons and five hundred men to appear at St. Georges, a town along the Chesapeake & Delaware Canal. To what end, no one knows. A journal entry from neighboring judge James McCauley several years earlier suggests another project. "This day is a day of rain. After dinner went to see John Randel Jr. Found him and wife very polite. Showed me a circular wheel in the center of the table with the degrees marked on it used in platting land. The paper is fastened to the wheel and a square slides along the edge of table to take the bearings or mark lines on the paper. It seemed to be an expeditory method of platting," the judge wrote.[115]

Johnston's assessment was that Randel squandered his fortune in "the prosecution of wild, chimerical schemes for self-aggrandizement." And as no countervailing account survives, Johnston's account is the one cited when Randel is mentioned in books or articles. One of Randel's descendants takes issue with the conclusion and counters with family stories. She recalls her great-grandfather's fondness for Randel. "He respected him for what he did. Thought that he was a visionary, trying to make the world a better place. That should be enough," says Lee Howard Vosters, who now lives in one of Randel's houses on Randalia Road. "He was eccentric, but my great-grandfather really liked him. Darwin and Audubon were in their time considered eccentric. In our family, eccentricity has never been a negative." She says Randel had a reputation for being very private—not a recluse, but not spending much time with people. "Maybe he was working on projects. He had an engaged mind."[116]

Johnston was correct about the squandered fortune, though. In 1849, Randel wrote to Clayton about finding work, saying Letitia did not want him to retire to their farm to "rust out." And a profusion of court documents chronicles the woes of debt. By the early 1850s, Letitia and Randel were seeking to sell or to protect what remained of their assets, and chancery records and testimony from lawsuits reveal their labyrinthine efforts. In 1852 they put their steam sawmill up for the sale: "One of the very best situations in the union for procuring Timber," read the advertisement. They owed money—about $14,000—to a neighbor, Jacob C. Howard, and drew up a trust in Letitia's name, with their son, John M. Randel, as trustee, giving Letitia much of their remaining land. "People often did that to hide assets from creditors," explains Hendrik Hartog. "This often gets misread by gender historians as women getting their own property. But it may have been a deal within the family to protect family assets. Which obviously didn't work in this case; they got caught." Letitia's was a modern trust, Hartog notes, because she retained the power to sell land and to revoke the trust; probably, he says, either she or the lawyer insisted on those terms or it was standard practice in Maryland at the time. Some of the land in the trust, however, was already mortgaged to Howard. And the language of the cases is complicated, contradictory, and unclear, because, as Hartog notes, a lot was going on outside the court: "Clearly both sides are negotiating and strategizing."[117]

If Randel and Letitia did design the trust to hide or protect their assets, that might explain an 1856 New York lawsuit in which Letitia sued Randel for $1,390 through her trustee, their son. Randel had borrowed the money from her and John, who was practicing medicine in Pennsylvania at the time. Letitia won. "It could be a way of scaring off creditors," Hartog notes. "Although it is unlikely that the trustee would sue the father under these circumstances, it is not impossible. It seems inefficient, and it is expensive." That lawsuit suggested that Randel and Letitia had perhaps separated. But the two appeared together again, in home and in court, well after the case.

Despite their legal machinations, the Randels' plight worsened. In May 1858, Cecil County seized their possessions because they had not paid Howard. "On Monday sheriff dispossessed John Randel the celebrated engineer who some years ago obtained a judgment against the Chesapeake and Delaware Canal Co. for $226,000. He has trifled it all away," Judge McCauley wrote. The sheriff's department listed the seized items in the *Cecil Whig*, including a family carriage, a buggy wagon, two cows, a wheat drill, a box, seven hogs, twelve pigs, one workstand, one piano, one bureau, one looking glass, a bedstead, a bed, clothing, and a clock. A few months later, on July 13, Letitia and Randel's only remaining child died. John M. Randel had contracted yellow fever when he had gone to Virginia several years earlier, responding to requests for help from local physicians. He may have been weakened by that illness, or he may have contracted another. There is no record of what killed him, at the age of twenty-seven.[118]

Although Randel and Letitia had lost their land, they did not stop fighting in the courts to get it back. In 1860 they appealed to the U.S. Supreme Court from the Maryland circuit court, where they had brought a case against Howard on February 18, 1859, that had been dismissed. They described themselves as citizens of New York. And they described coercion, which was not the standard language of a fraud suit. They argued that Howard had intimidated and threatened them. Randel testified that he at first refused to sign Howard's agreement but Letitia forced him to because she was scared for her life. The Supreme Court could make little sense of the case, writing that "the statements of this bill are vague and uncertain, frequently argumentative, and very rarely plain and direct. The whole bill lacks definitiveness. Agreements, friendly arrangements understandings, and fraudulent devices are spoken of, but the character of the agreements and the nature of the devices we do not learn . . . Are the complainants in a situation to enforce the trust, if one is established? We think not." Cecil County chancery records show that Letitia kept fighting Howard well after Randel's death. She was apparently unsuccessful and spent her later life living with Randel's nephew in New Jersey.[119]

The Randels and the Howards eventually put aside their differences and married into each other's families: Lee Vosters's nephew is John Randall Howard, and Vosters herself was nearly

Randelia today. Photograph by the author.

named Letitia. Randel's small fieldstone home has been extended and modernized, but the original main room endures. It has a low ceiling, a trap door leading to a sleeping loft, and a large fireplace, over which Randel would hang his muzzle-loading gun. Ernest Hemingway once sat in the small cozy room, visiting as a friend of Vosters's father, Polk Steele Howard, who hunted and fished and knew Hemingway from Cuba. Vosters used to keep a dairy cattle farm, an asparagus farm, and the second largest sheep farm in Maryland on Randelia. Now she keeps blue-sheened guinea hens, corgis, and horses. Tall hedges surround empty fields and line the road. The land of Randelia, although no longer so called, remains green farmland. Thickets and woods hug the waterfront. Ducks still abound.

V

In Which *Mannahatta* Lifts Off

Columbus Day morning 2003 was clear, sunny, cool, not cold. Eric W. Sanderson was walking east on 42nd Street, heading toward the assembling annual parade. He carried a small GPS unit, not much larger than a cell phone. Wearing blue jeans and a light coat and shouldering a heavy green backpack, Sanderson looked ready for a hike in the woods—which he was. As he approached Fifth Avenue, Sanderson paused. "So probably here was oak and chestnut forest," he said, pointing at Bryant Park. He then gazed southeast at the swell of Murray Hill. The Murray family buildings and the shaded rise of the hill can be seen on Randel's 1811 map and on farm map 18. "There is a legend that Mrs. Murray entertained the British general to try to delay him," Sanderson said. "What seems more likely is that the general was under orders to stop at Murray Hill. But if he had just walked down the hill and blocked this road, that would have been it. Instead, a mile away, the American troops got away again." He stood a few seconds, entranced by what his mind's eye could see.

On Fifth Avenue, bagpipes and motorcycles surged. Paradegoers brandished Italian and American flags. "A lot of primates," Sanderson noted. "Needless to say, it would have been a lot quieter." He turned north onto the west side of Fifth and descended into the cool air of a steep 30-foot-deep ravine. Brooks and birds were soon the only sounds beneath beech and hemlock trees. Sanderson began walking the valleys, hills, and forests of an earlier, less noisy Manhattan.

A quarter of a mile later, Sanderson stopped to listen to a stream coursing down a slope under what had become the J. R. Cigar Store on the northwest corner of Fifth Avenue and 46th Street. (The store is no longer there.) He crossed it—"probably a very nice little stream, babbling brook as it were"—and turned west, down Diamond and Jewelry Way on 47th Street, past more hemlocks, some wild turkeys, a spring (the source of the stream), and a small pond, ideal for duck and beaver. The block became the heart of the diamond trade in the early twentieth century, as merchants moved north with the unfolding grid; the diamond district was originally based in the Bowery, near Canal Street. As Sanderson gazed at the pond, a woman pushed past him to gaze at diamonds in the window of a jewelry store. She too was peering into the past, across millions of years when, miles beneath the crust, carbon was pulverized into tightly bound cubed lattices and then pushed upward in eruptions of magma. The carbon crystals cooled and sat until they were cut out, cut up, and caratized. The woman may not have appreciated the reach of time as she looked at the flickering rocks. Eons are hard to fathom. Even the changes across four hundred years are hard to imagine. Nevertheless, Sanderson has devoted more than a decade to getting New Yorkers—and, by extension, city dwellers anywhere—to see the past alive in the present.

Sanderson is the author of *Mannahatta: A Natural History of New York*, a best-selling book many New Yorkers and others came to know well during the four hundredth anniversary of Henry Hudson's arrival in his flyboat, *Half Moon*. In the book and an ongoing web-based project, Sanderson presents a vision of Manhattan in 1609, the year Hudson arrived and the year marking the beginning of European transformations of the land, of species composition, and of native people's way of life. Informed by scientific information and imagination, *Mannahatta* (the Lenape's word for their "island of many hills") is the first computer model seeking to render a four-hundred-year-old landscape in an approximation of its full ecological splendor. Sanderson hopes *Mannahatta* will enable New Yorkers to layer transparencies of the past atop the present, and to imagine future layers, future possibilities.

To create *Mannahatta*, Sanderson had to lift the grid, which connected him to the man who set it down. Randel's topographical data proved crucial to Sanderson's ecological restoration. Sanderson loves thinking about storms, trees toppling to create space for saplings, marshes rich with birds, fry, and mollusks. Randel enjoyed eating oysters but was plagued by thickets, storms, and mosquitoes swarming from the island's profusion of wetlands and marshes. Across the expanse of two centuries, the two men have met, perfect foils in their double vision. One walked a verdant island, seeing right angles and the streets of a great invisible city. The other walks a great gray gridded metropolis, seeing green hills.

REMOVED FROM A WILDERNESS CONDITION

The idea of ecological restoration—of recreating a landscape—arose in the United States at the tail end of Randel's era, during the full throttle of railroad construction and industrialization. As the New York City grid extended up Manhattan and the federal land ordinance grid extended across the West, as canals and then railroads linked states and the vast interior, the American relationship to land and nature changed. Infrastructure transformed landscapes and fostered greater economic security; the nation grew less skittish about its future. Americans increasingly came to reflect on the transformation of familiar landscapes and on the loss of wildlife. They had hunted, farmed, and "improved" the land for more than two centuries, and the forces making settlers vulnerable—the chaotic forces Gouverneur Morris feared in 1806 might "unhinge the intellect"—were no longer so viscerally troubling. By the mid-nineteenth century, more than 60 percent of New England's old-growth forests had been felled and many animals were long gone. Americans were becoming nostalgic, appreciating wild nature, wild land—the particular American landscape—now that they no longer felt threatened by it and the possibility of depleting the nation's ample resources loomed. "Although there were a few exceptions, American frontiersmen rarely judged wilderness with criteria other than the utilitarian or spoke of their relation to it in other than a military metaphor. It was their children and grandchildren, removed from a wilderness condition, who began to sense its ethical and aesthetic values," writes historian Roderick Frazier Nash.[1]

The shifting sensibilities were captured in the early and mid-nineteenth-century by authors such as Washington Irving, James Fenimore Cooper, and Henry David Thoreau, the philosopher Ralph Waldo Emerson, and the painters of the Hudson River School, among them Asher B. Durand, who had worked with the master engraver Peter Maverick during the time Randel designed his 1821 map. These writers and artists depicted nature as sublime and essential to the health of the human spirit, and they celebrated the distinctive American landscape. Central to their Romantic or transcendental views was a conviction that nature existed in a harmonious equilibrium and that humankind could disrupt the balance of nature and increasingly did so through the use of technologies. The railroad rapidly became "the machine in the garden," as literary critic Leo Marx calls it. A tension between industrial progress and the preservation of essential nature—for some the pastoral, for others the wild—emerged.[2]

How fully, and to what end, humans could alter nature became more widely clear to Americans in 1864, when scholar, linguist, and diplomat George Perkins Marsh published *Man and Nature; or Physical Geography as Modified by Human Action*. Marsh's popular book used science and history to argue that humans unintentionally as well as intentionally destroyed natural processes, creating disastrous consequences for themselves. They wrought devastating floods and lethal droughts: ". . . man is everywhere a disturbing agent. Wherever he plants his foot, the harmonies of nature are turned to discords."[3]

The idea that changing the landscape changed human experience was not new to many Americans. "Most New England naturalists agreed by the 1790s that deforestation and agricultural cultivation had the effect of warming and drying the soil, making the surface of the land hotter in summer and colder in winter. Temperatures in general fluctuated more widely without the moderating effects of the forest canopy to shade the ground and protect it from winds," notes University of Wisconsin environmental historian William Cronon. But Marsh's book reached an extensive and diverse audience: it was read not only by naturalists, scientists, and historians but by the general public. And Marsh's argument was global, rooted in resonant historical examples. Environmental destruction could do more than change local weather—it could destroy civilizations. He linked deforestation and erosion to the fall of the Roman Empire and to the emergence of vast deserts.[4]

Marsh's detailed, rigorous, highly footnoted, and digressive treatise advocated for human intervention on nature's behalf for the mutual benefit of man and nature. Having observed efforts to replant forests and stabilize mountainsides in Europe, Marsh argued that nature needed humans in order to heal. Man "is to become a co-worker with nature in the reconstruction of the damaged fabric which the negligence or the wantonness of former lodgers has rendered untenantable," he urged. His call for environmental restoration appears to be the first made in America.[5]

It was to be many decades—indeed, more than half a century—

before ecological restoration was formally pursued as a discipline and before the full significance of Marsh's arguments was appreciated, but Marsh's influence was far-reaching in his own time as well. Some scholars credit him with launching the U.S. conservation movement, contributing to the mindset that led to the establishment of the first national and state parks. And although Marsh did not advocate for parks in urban areas, his ideas about the human relationship with nature resonated in aspects of Frederick Law Olmsted's projects. Olmsted repaired damaged land in Boston's fens and along the Muddy River, reclaiming a polluted urban wasteland. As scholar Anne Whiston Spirn notes, the project seems to be the first effort in any country to "*construct* a wetland." Olmsted built water purification and flood control into this wild-looking, nonpastoral landscape. By dredging, draining, recontouring, and planting native and nonnative plants, he made an ecosystem. "The thing is to make it appear that we found this body of water and its shores and have done nothing to them," he wrote.[6]

With Central Park, which predated the Boston Fens and Riverway by several decades, Olmsted was not seeking to construct a sense of wild Manhattan. Although he advocated for the protection of Yosemite and for the preservation of nature in untrammeled form in other places as he worked on Central Park, he wanted a pastoral, rural experience for New York City dwellers. Central Park, as noted in Chapter III, is landscape architecture—an artfully engineered Romantic setting, an idealized, human-friendly nature.

The *Mannahatta* idea is a Central Park for its time. It is a form of restoration ecology, not landscape architecture. It is virtual, physically aspirational, not actual. Yet both the park and *Mannahatta* celebrate the value of nature in an urban environment. Both seek to connect people to a lost or inaccessible landscape. Both remove the grid and by so doing criticize the grid for evicting nature from the city. Both are informed by a Romantic sensibility.

Both posit a stark opposition between "nature" and "city," the garden versus the machine. Both use the cutting-edge technologies of their time to create and evoke "nature."

On-the-Ground Truth

Maps, particularly historical maps, stir the individual imagination of viewers. The mapmaker selected what was important to him or to his sponsors. Europeans portraying the New World wanted enticing maps, as did Simeon DeWitt and Randel. But the cartographer's chosen details can elicit vivid and unexpected images in the viewer. Personal experiences converge with seemingly universal referents; the viewer's idea of mountains (for some the Berkshires, for others the Himalayas), the viewer's expectations, hopes, prejudices, sense of way-finding, and fantasies all come to bear. Which is why beautifully made and historical maps are such treasured objects—treasured as much for what they evoke as for their artistry.

Mannahatta was inspired by just such a map. Sanderson moved to New York City from California in 1998 to work as a landscape ecologist for the Wildlife Conservation Society at the Bronx Zoo. Part of his job would be to create computer maps of habitat, of resources, of animal territories. One weekend a few months after moving, he visited the Strand Bookstore and discovered *Manhattan in Maps*. He starting paging his way forward from the lost Maggiolo Map of 1527—the first depiction of the area that was to become New York City. Whereas Reuben Rose-Redwood had been captivated by Randel's farm map 27 from 1819, Sanderson stopped short several pages and decades earlier, at the British Headquarters Map, circa 1782.

The map, drawn at a scale of 1 inch to 800 feet, had not been reproduced since 1900. Robert T. Augustyn and Paul E. Cohen, the authors of *Manhattan in Maps*, were the first to publish the British Headquarters Map in full color, the first to introduce it to modern audiences. The map portrays the many hills of Manhattan shaded, inked in beautiful relief so they seem to rise above the page. As the authors note, occupation of the island between 1776 and 1783 meant that British surveyors, engineers, and "the cream of English cartographic talent" were in long-term residence; as a result, "New York, which was one of the most poorly mapped American cities before the war, became by its end the most thoroughly mapped urban area of the United States." Based on the landscape presented in the map, Augustyn and Cohen describe what a "rich and varied visual experience a carriage ride" around the island would have offered during that era.[7]

The images the map conjured for Sanderson were at less a remove than a carriage ride. The meticulously rendered landscape, woven with so many waterways that "island of many streams" would have been an equally apt Lenape name, conjured images of vegetation, insects, amphibians, reptiles, mammals, and birds. As a landscape ecologist, Sanderson knew topographical information and details about watercourses could yield clues about what the island's ecosystems had been before they were covered in cobblestone, asphalt, and concrete. "Why is some place forest, some place wetland? It is the geomorphology, the topography, the hydrology, the shoreline," he said. Perhaps, he thought, relevant information from the map might be combined with historical and contemporary records of flora, fauna, climate, and soil to understand—or at least reasonably hypothesize about—what used to live where on the island. Sanderson bought the book. And the more Sanderson examined the British Headquarters Map, the more he became convinced it could unlock the island's ecological past. Maybe he could create a computer model and produce historical ecological images that could, ideally, show any New Yorker or tourist what the corner he or she was standing on used to look like. Perhaps he could induce New Yorkers to "identify with the nature that used to be here as much as they identify with Times Square today," he mused. "And when they think of New York, they don't just think of Times Square and taxis, but they think of forests and wetlands and deer and wildlife." He paused. "Maybe not deer; there are too many of them. Bears and elk or something."[8]

The first step to building the model would entail scanning the map into his computer so he could manipulate it. To scan it, he had to slice apart the pages. This act proved to be one of the most difficult of the entire project. It took him six months. "It is hard to cut open a book. Was it worth it to cut the binding of a fifty-dollar book?" he asked. Finally, one Sunday afternoon, "I was like, goddammit, I am going to do it! I got out my X-Acto knife and cut it open and scanned the map and assembled it with a program I had."

Sanderson then planned to position the scanned map over a contemporary digital city map, and for that he needed control points. In the same way that Rose-Redwood required the datum so he could relate the height of a hill in Randel's terms to the height of that same hill in today's terms, Sanderson needed to correlate landmarks on the British Headquarters Map with the GPS coordinates of those landmarks today. Such landmarks are like pins holding together the back and front of an unfinished garment. The process of connecting one map to another through those points is called geo-referencing or—a lovelier term of art—ground-truthing. In 2001, Sanderson and a crew of two began traveling the island to look for some fifty landmarks that had probably survived since 1782: the hill in Highbridge Park where Fort George had stood, old walls around Fort Tryon, Mount Morris in Marcus Garvey Park, the steps of Trinity Church at Broadway and Wall Street, St. Paul's Chapel on Broadway and Fulton. Sanderson's team consisted of his newborn son, Everett, and his wife, Han-Yu Hung, a horticulturalist and educator at the New York Botanical Garden. "The field crew is in good condition and high spirits despite the early hour," Sanderson narrated in one of the home videos taken during the early fieldwork. "It is so different from being on a field trip in Gabon. You can stop and get a cappuccino," he noted. "I sat in a café downtown with the map and looked outside and tried to reconcile this in my head. This pond is gone. That hill is gone. That rise is gone. Remarkable how much it has changed. It blows your mind." About twenty-five landmarks were shared by the two maps after those early forays.

But when Sanderson tried to align the British Headquarters Map with a contemporary one, affixing the two together at the landmarks in common, "there was a big problem." He found that the 1782 map was peculiarly out of register with his current one. Overall, the discrepancy was about 100 meters, which didn't bother him greatly. He felt he could work with that degree of error, and over time he reduced it to an average of plus or minus 40 meters. For a surveyor, a discrepancy of 100 or 40 meters would be akin to looking at a 3-D movie without the funky glasses: blurred edges, no focus, distraction, a powerful headache. For a virtual ecological reconstruction that was to be in large part inference and artistry, that fuzziness was acceptable. Much more troubling to Sanderson was the fact that the two islands were quite dissimilar to the south. "It looked fine for most of Manhattan, but the lower tip of [the British Headquarters Map] jutted out into the Hudson River." After days of consternation and frustration, he noticed a faint dotted line on the old map. He returned to his computer scans and shifted pieces, reassembling the two southernmost sections along the faint line. The 3-foot-by-10.5-foot map had been misassembled in the National Archives in London, where it is kept and where Sanderson visited it, still misaligned, in 2001 and in 2006.

Once the British Headquarters Map overlapped the island of today, Sanderson could start to add the components of lost ecosystems. But which ones? Those present in 1782, during the British occupation? Those present at the birth of the modern city in 1811? Sanderson needed, as all restoration ecologists do, a baseline. "People think of the Commissioners' Plan as being the big thing," he reflected. "I am not so sure about that. If you think about biological changes, when the biggest changes occur, I think it was when the Dutch came in the 1620s to settle and introduced agriculture, pigs and cows. They just let them loose in the woods. Those things have a massive effect on ecosystems in terms of competition, introduced disease, and pests. And on plants. Those were all huge things. By the time you get to actually laying down roads and buildings there had already been so many modifications."

During the seventeenth century, the Lenape relationship with and impact on the island—burning, hunting, gathering—ended as well. "The Native Americans abandoned Manhattan after a really nasty war," Sanderson noted. Under the combative and harsh policies of governor Willem Kieft, the Dutch of New Amsterdam were exhorted to fight the Lenape, a conflict triggered not solely by Kieft's reportedly blustering and bloodthirsty nature, but by the colony's decreased reliance on the tribe for trade and increased demand for tribal land. (Kieft was ultimately recalled to Holland and replaced in 1647 by Peter Stuyvesant.)

Sanderson eventually settled on 1609. The year represented the liminal moment between Native American and European effects on the land. It also offered an evocative moment in the city's history. Most New Yorkers, history buffs or not, know of Henry Hudson, and the four hundredth anniversary of Hudson's arrival was only a few years off. Sanderson had a built-in deadline and the potential for citywide engagement if he linked the project to the quatracentennial. He was ready to recreate the ecosystems of 1609, to bring Mannahatta back to life.

"Landscapes are among one of the most intricate and fascinating historical documents that we have, and too few see it that way. Most people take the landscape for granted," commented William Cronon when Sanderson embarked on the project. "But for people who have eyes to see, you can see that the entire past exists as ghost layers. It is palimpsest. You can peer down through those different descriptions of the record of human dwelling and all the natural systems and creatures. Arguably, what better place to do that than Manhattan, where people don't think nature exists?"

Measure for Countermeasure

The surveyors who set Americans along their paths to ownership and nationhood, who staked out grids small and large, who sought the shape of the earth, unwittingly provided an antidote to their own actions. For centuries their documents have provided crucial legal information about provenance and property lines. That legal value persists, but another value has emerged in tandem: surveyors' records and maps provide vital information about former landscapes, permitting ecological rewind—permitting us to assess what shape the earth was and is in. Many historical and restoration ecologists celebrate their luck in that they have several centuries of survey data containing details about trees, vegetation, water, and soil in locations that can be identified today. The land ordinance of 1785 led to the creation of more than 6,500 volumes of survey notes kept by the U.S. General Land Office (which was formed in 1812 to manage the surveys and settlement) and more than 100,000 township maps. Additional state and town records, such as those Randel made, supplement the wealth of federal records. A few ecologists began using these records in the 1920s to understand the land's history, particularly in Ohio and Indiana. Their efforts, which culminated in manuscript maps and dissertations, were little known, and had largely wound down by the 1960s. But the emerging field of GIS revitalized the practice.

In Randel's day, a typical map showed the general features of landscape and property. Cadastral or census information might be printed around the edges, as in Randel's 1821 map of New York City and the northeastern states, or in an accompanying pamphlet. Or a map might be thematic, portraying one kind of information, as in William Smith's 1815 geological map of Great Britain. The hallmark of a GIS map is multiple layers of information. For that reason, physician John Snow's famous 1854 map tracing cholera cases in London to public water pumps is credited as a kind of GIS precursor: his map revealed a connection between contamination, exposure, and disease. In essence, GIS enables mapmakers to layer various kinds of information on a landscape or geographic region and present it in different combinations. These associations ideally give rise to insights about an area or an issue. A GIS map might show a city's Zip Code zones and then add strata of data about population, air pollution, disease, and income. It might combine vegetation cover with average mean temperature. The overlaps might yield connections significant for policy or research, such as showing an association between the number of asthma cases and proximity to bus depots.

GIS arose in the late 1950s and early 1960s in different settings for different ends, including the depiction of land use, transportation patterns, census data, ocean depths, urban planning, and natural resource management. Although it was not initially used for historical recreation, historians and ecologists are increasingly employing it in creative ways. Among the first historical ecologists to use nineteenth-century surveying records in such a model was David J. Mladenoff of the University of Wisconsin. The idea of documenting the state's original landscapes had been considered for a long time, "because Wisconsin was so heavily transformed," Mladenoff explains, and because the Wisconsin Historical Society in Madison held the field books and maps of about one hundred surveyors who had walked and staked out the state's townships and ranges. Mladenoff started a project to do so in the 1970s, but abandoned it because the GIS was too cumbersome. In 1994 he tried again. Working with GIS expert Ted Sickley and an "army of undergrads," Mladenoff began recording every detail the surveyors noted or marked on maps as they set markers or scratched signs into bark every half-mile. The marked trees were called bearing trees or witness trees.

The surveyors' accounts are virtually identical to those Randel, Charles Brodhead, and others made in upstate New York for Simeon DeWitt. "Land first half level prairie last half hilly stony timber oak," reads an October 1832 entry by John H. Mullett. The entries also show surveyors' woes to be universal: "During four consecutive weeks there was not a dry garment in the party, day or night . . . we were constantly surrounded and as constantly excoriated by swarms or rather clouds of mosquitoes, and still more troublesome insects; and consider further that we were all the while confined to a line; and consequently had no choice of ground . . . and you can form some idea of our suffering condition. I contracted to execute this work at ten dollars per mile . . . but would not again, after a lifetime of experience in the field, and a great fondness for camp life, enter upon the same, or similar survey, at any price whatsoever."[9]

For five years Mladenoff and his team entered the information into a database. When they were finished, the records enabled them to create statewide maps of soil and vegetation, of prairie and woodland. Unlike contemporary satellite images, which can reveal only whether vegetation is evergreen or deciduous, the images created with the nineteenth-century survey data provide species-level information, identifying white cedar, tamarack, spruce, black ash, white pine, sugar maple, yellow birch. With such fine resolution at their fingertips, the team discovered important details about the older landscape. "People in Wisconsin used to think an area of conifer forest along Lake Superior was boreal forest," notes Mladenoff, meaning that one of the predominant tree species would have been balsam fir. "Actually, it was predominantly white pine. That was new information for them and led them to reevaluate what they were doing." The Wisconsin Department of Natural Resources, which helped fund the project, uses the data to make management decisions and to prioritize land for protection. The Nature Conservancy has used the maps for a restoration project.

Mladenoff knows of similar statewide efforts in Michigan and of many other historical ecologists using surveyors' data. But the most ambitious project he is familiar with is Charles V. Cogbill's. It transcends states or a particular forest.

Since the early 1980s, Cogbill, a freelance ecologist based in Vermont, has been traveling the Northeast gathering surveying and planning information for 1,350 early towns in nine states. He recalls learning about an old map for a town in New Hampshire where he was studying a forest, tracking down that map, and having "a Eureka moment, if you will." The trees standing in the corner of each lot—the witness trees—were labeled with species names. Cogbill realized he could recreate the exact location of specific types of trees all over the Northeast with such town maps, noting valuable information about the early landscape and its flora. Since that cartographic encounter, Cogbill has compiled information about well over 200,000 witness trees. He has used only one Randel survey in his set so far, which provided information for part of the Onondaga Salt Springs Reservation—although "DeWitt is on my horizon, to see if there are some geographic holes that I have that he might be able to fill in." Every data set, of course, has its caveats. Cogbill says bias exists in the surveyors' reckoning of what the land was best suited for—in how they described its utility. Tree names are not standardized and not always easily recognizable, and so Cogbill has developed a lexicon for state-specific colloquialisms. The project is ongoing: "If you could tell me what Cockburn's rose bloom was, I would be forever grateful." And sometimes surveyors recorded only genus, not species. Yet, Cogbill says, "almost every surveyor was a very good natural historian. They were very good botanists, I am convinced. The interesting thing is how consistent and accurate they really were."

Cogbill's findings have altered ecologists' understanding of the

Northeast's pre-European landscape. Most previous studies took the remaining 1 percent of old-growth New England forest as a baseline, extrapolating conclusions about the original forest from those few stands of remnant trees. But Cogbill and his colleagues contend that those patches survived precisely because they were *atypical*. Using them to inform understanding about the original landscape is akin to looking for keys under a lamppost. He has shown how pivotal topography is for the balance of species, and that pine, hemlock, and chestnut were not as extensive as ecologists previously thought. Instead, beech and oak were more dominant: beech to the north, oak to the south. The findings have implications for how GIS programs model the forest, for management, and for restoration. "We are just scratching the surface," Cogbill says. "The material is voluminous, and the types of information are multidisciplinary."

Science with a Dash of Poetry

No witness trees existed for Mannahatta. The surveyors creating the British Headquarters Map were interested in forests for fuel and the features of the land for military maneuvers. To find information, such as Cogbill's, about what types of presettlement forests grew on the island, Sanderson had to infer soil types from geology and hydrology. And he extrapolated tree species from those inferences as well as from elevation and slope. Some trees like more dampness, some less; some like to live higher up, some lower. Some are fine with sandy soil, others not. Some prefer a southern exposure; for others, north or east or west is best.

Sanderson felt confident about the hydrology, because he had combined the British Headquarters Map with Egbert Ludovicus Viele's famous 1859 water map. Viele, a civil engineer and rival of Frederick Law Olmsted for the Central Park design, mapped the island's water systems to make the case that diseases such as cholera and malaria, still rampant in mid-nineteenth-century New York, could be curbed by drainage and grading. "It is a well established fact that the principal cause of fever is a humid miasmatic state of the atmosphere, produced by the presence of an excess of moisture in the ground, from which poisonous exhalations constantly arise, vitiating the purer air and carrying into the system of those who inhale it a virus," he wrote in an 1865 pamphlet, *The Topography and Hydrology of New York*. "Let us hope the time is coming when we shall do some credit to the higher intelligence and broader philanthropy which characterize the age in which we live, and shall adopt those measures which are so clear and so imperatively necessary . . . In this money-making money-wasting generation, let us not be deaf to the lessons of the past."[10]

Some of Viele's recommendations did come to pass. But water is not easily dominated. Today, when flooding occurs on a construction site or when a foundation sinks, engineers and architects consult Viele's gemlike green-and-blue map. There they often see the thin green line of a stream gone underground. Unbeknown to many of them, they are largely relying on the research of another cartographer. Viele consulted Randel's farm maps to determine where the island's watery features lay.[11]

Inevitably, Sanderson came to rely on Randel too. He needed precise elevations, not evocative shadings capturing a hill, a bigger hill, a really big hill. Unless he knew heights, he could not plant tree species or other flora in the right places on his map. And without the vegetation characterizing the ecosystem, he could not add the animals associated with those particular ecosystems. Sanderson enlisted the aid of a Columbia University biology student. During the summer of 2003, just as Rose-Redwood was finishing his first digital elevation model and readying it for publication, William T. Bean, who is now a doctoral student in ecology at the University of California at Berkeley, started gathering elevation

Sanitary and Topographical Map of the City and Island of New York by Egbert L. Viele. Courtesy of the David Rumsey Map Collection, www.davidrumsey.com.

data from Randel's records. Settling in with the field books at the New-York Historical Society, Bean recalls, "was exciting because it was the first real primary data that I was working with, the first time it seemed like I was doing something that someone hadn't done before. And that was really thrilling." But it wasn't clear that it was going to work. "There was the risk that I was collecting all this data and that we weren't going to get anything from it. It was not well organized, and incomplete."

Bean could discern distance and temperature, but he was not sure how to decipher some of the other figures. It wasn't until he went home for winter break and showed the numbers to his father, a computer programmer, that it all came together. With his father's help, Bean realized that Randel had been recording angles of ascent and descent for every rod or chain measure he made. So a cross street and an avenue had complete and, most importantly, decipherable elevation profiles—angles of rise or fall every 10, 30, or 50 feet. Bean first tried using these elevation data in his senior

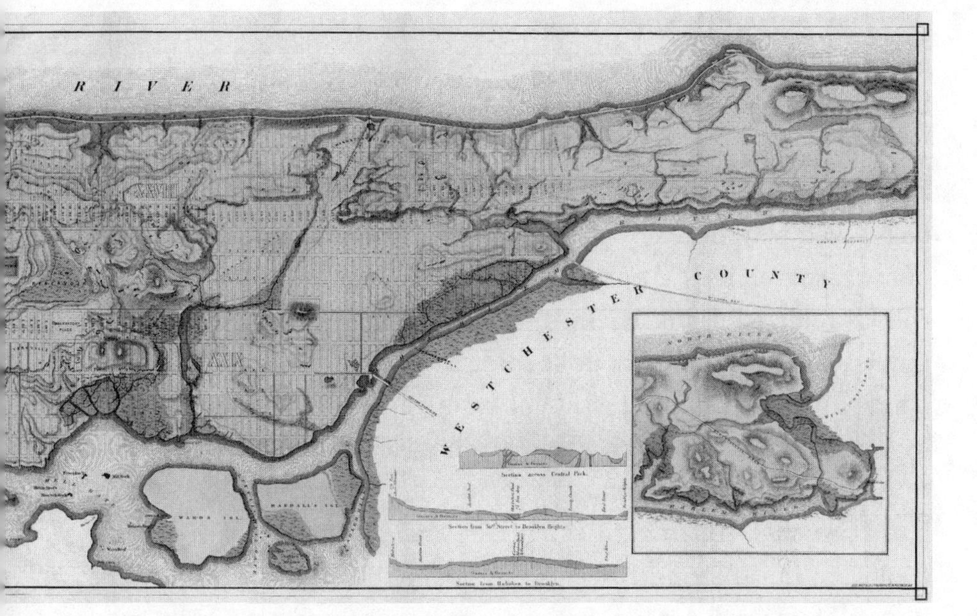

paper, on the burning patterns of the Lenape in Harlem plains—elevations Randel had gathered when he lived in Harlem, tending his garden in a former grassland.

To create their elevation model for *Mannahatta*, Sanderson and Bean printed out a version of the British Headquarters Map and added all the elevation data they had collected: Randel's farm map elevations and field book data, elevations recorded on Viele's water map, and elevation figures from Sanderson's and others' GPS recordings at enduring landmarks and rocks—more than 7,000 points, including the initial 25 Sanderson and his family had collected. Then they correlated some of the elevation data to hill shadings on the British Headquarters Map, drawing contour lines by hand, which took nearly four months. "That was the only way we could integrate the map and an elevation model," Sanderson says. Bean entered the fruits of their effort into a software program called ArcGIS, and *Mannahatta's* digital elevation model was born.

Bean currently studies the biology of giant kangaroo rats, endangered burrowing rodents that have lost 99 percent of their habitat. He has spent months at Carrizo Plain National Monument, 100 miles or so from Los Angeles, figuring out where the kangaroo rats thrive so he can predict the best habitat to preserve. Randel has stayed with him across the hot empty desert. "A large portion of my research is fieldwork, and I do keep a field notebook. I don't think I realized it, but having access to Randel's notes had a huge impact on how I think about keeping records. Something I thought about constantly while I was digitizing his notebooks was the fact that he must have had no idea (1) what New York would turn out to be, and (2) how his surveys would be used (e.g. the *Mannahatta* project)," Bean reflected in a recent e-mail. "I call myself an ecologist, but of course ecology is deeply entwined with natural history. And while I hope that my research is important right now, I also suspect that what will be valuable one hundred years from now is not some computer model that I developed for kangaroo rats, but my notebooks—the natural history that I've recorded at this time, in this place. That is, I really think ecologists are historians of the land. I hadn't realized how much I believe that, and how much of that must be due to Randel."

And a Raye as Great as Foure Men

To the skeleton of *Mannahatta*—to the elevation, soil, hydrology, and to some solid temperature data—Sanderson now introduced species and flux. Many travelers and residents had chronicled New York City's natural history. "The Land they told vs were as pleasant with Grasse and Flowers and goodly Trees, as euer they had seene, and very sweet smalls came from them," wrote Robert Juet, Henry Hudson's first mate, in September 1609, amid his many notations about magnetic variation and how far off true north the compass needle strayed. "So wee weighed and went in, and rode in fiue fathoms, Ozie ground, and saw many Salmons, Mullets, and Rayes, very great. The height is 40. degrees, 30. minutes . . . Then our Boate went on Land with our Net to Fish, and caught ten great Mullets, of a foot and halfe long a peece, and a Ray as great as foure men could hale into the ship." Juet noted chestnut and walnut trees, pumpkins, grapes, and bear and otter skins. A director of the East India Company recorded Lenape words for many animals, so Sanderson also knew that the island had deer, elk, cougars, skunks, turkeys, partridges, loads of toads, and tons of turtles. One of his favorite Dutch records read, "The most wonderful are the dreadful frogs."[12]

Some early chroniclers were scientific in their orientation, using Linnaean classification and avidly corresponding with other natural historians of their era. Others were less exact. But all captured abundance. "Boats crossing the bay were escorted by schools of playful whales, seals and porpoises. Twelve-inch oysters and six-foot lobsters crowded offshore waters, and so many fish thrived in streams and ponds that they could be taken by hand. Woods and tidal marshlands teemed with bears, wolves, foxes, raccoons, otters, beavers, quail, partridge, forty-pound wild turkeys, doves 'so numerous that the light can hardly be discerned where they fly,' and countless deer 'feeding or gamboling or resting in the shades in full view,'" write historians Edwin G. Burrows and Mike Wallace in their review of those early accounts."[13]

While Sanderson compiled accounts of plants and animals, he reached out to several dozen archaeologists, anthropologists, historians, paleobotanists, biodiversity mavens, herpetologists, invertebrate experts, foresters, and botanists, who contributed their wealth of knowledge about the island's physical, biological, and human history. Many scientists Sanderson spoke with in those early days were intrigued but skeptical. "The project as an idea is a very exciting and excellent one," mused seismologist Klaus Jacob at the time. "My concern is, can he really extract enough hard data? He has a good model into which he feeds all the physical environment, and he thinks he can reconstruct total ecology from spare data. That is a tough nut to crack." Sanderson's fusing of a rigorous scientific approach with scant data might produce a historical myth—of which plenty already abounded.

This is the challenge of restoration ecology: the distance between what we truly know about the past and what we want for the future, tempered by our truly inadequate, ever-shifting ecological understanding. One of the earliest, if not the first, U.S. restoration projects is now seventy-five years old. Curtis Prairie—which John Curtis, Aldo Leopold, and others decided to restore in the 1930s—remains a work in progress. Controlled burns, which help grass seeds germinate and keep woody plants from thriving, are not dissuading enterprising shrubs and invasive species. The webs of interactions between members of a community—from the microorganisms populating the soil and root systems to the relationship between a plant, a pollinator, and a particular bird—are dizzyingly intricate, still mysterious today, when the planet has been so mapped and scrutinized. As E. O. Wilson notes in *Biophilia*, ". . . the naturalist's journey has only begun and for all intents and purposes will go on forever . . . it is possible to spend a lifetime in a magellenic voyage around the trunk of a single tree."[14]

And when Leopold started his prairie restoration, the ecological paradigm was balance of nature and equilibrium. The idea that the Romantics and the transcendentalists, that Marsh and his cohorts honored seemed to find scientific support with the work of a biologist named Frederic Clements. At the turn of the century, Clements had devised a technique that allowed him to show that plant communities moved through distinct phases, called stages of succession, until they reached a stable or climax state. Imagine a meadow first colonized by shrubs and then filling into forest. Clements's breakthrough technique, which did much to establish ecology as a modern quantitative science and which remains the predominant field technique today, was the grid. By plotting a study site on a grid and counting and characterizing species within 1-meter-square frames called quadrats, the technique made ecology mathematical, statistically robust. The harmonious proportions of the mathematical grid seemed to capture a harmony of a kind: the balance of nature. Quadrats revealed to Clements and other researchers that over time, ecosystems reached a state of equilibrium—an endpoint, a climax, a mature stasis.[15]

That endpoint vanished in the mid- to late twentieth century. In the 1950s, John Curtis, R. H. Whittaker, and other biologists showed that there was no equilibrium. Individual plants were not working together to create some preordained community; they were all out for themselves. This idea had been put forth in Clements's time by rival botanist Henry Gleason, but during that era the scientific community was not ready for it—flux was anomaly, equilibrium was the dominant paradigm. By the 1970s researchers were building on the work of Curtis, Whittaker, Gleason, and others, characterizing ecosystems as undergoing constant disturbance, as being "chaotic" and, then, "complex." Chaos has no order, whereas complexity suggests, as environmental historian Donald Worster points out, that there "might not be any large overarching order inherent in nature, but there was plenty of evidence of conditions of change giving way to those of order, of order dissolving into change." In short, nothing stays the same, ecosystems included. Although the ideas of equilibrium and of a climax community endure—they remain central to many people's thinking about wild nature—and have their place in biology, the ideas of flux and disturbance are just as essential.

Today Clements's quadrats capture flux, whether in the rocky intertidal zone of Prince William Sound or on a survey of botanical diversity deep in the Amazon. Ecological GIS programs model and incorporate that flux as well—a tree fall, a lightning strike, hailstorms, hurricanes, fire.

SANDERSON ACKNOWLEDGES ALL of *Mannahatta*'s scientific shortcomings. The information he amassed with the help of Jacob and many others and through his reading was, as all had anticipated, uneven. Completeness in the species lists: improbable. Completeness in the Lenape records: improbable. Accuracy in the computer models of flux, plant succession, and human impact: highly unlikely. "So what I say is that we use science to the extent that the science can help us. And then we admit when we don't know," Sanderson says. "But we don't have to be afraid because we don't know. I am not saying it is going to be easy. And I am saying maybe it is crazy. But I totally think it is worth trying."[16]

Mannahatta has brought engagement with restoration and historical imaginings to a wide audience because of its storytelling, scale, format, ambitious artistry, and because it explicitly muses about the future. The centerpiece of the project and the book are several stunning computer renditions by artist Markley Boyer of different areas of the island. Sanderson and his team's years of research and software tweaking and orchestration and Boyer's graphical talents culminated in sweeping bird's-eye views of a lush, verdant isle thick with forests, woven throughout with blue streams, home to bears, eagles, and egrets. Photographs taken above the Amazon rainforest seem the most apt modern comparison. Informing the images are Sanderson's Manhattan tallies: the island had 21 lakes or ponds, 66 miles of streams, 10,000 or so species (not including insects, molds, mosses, and microorganisms), and 55 ecological communities—among them sandy beaches, eelgrass meadows, red-maple hardwood swamps, vernal pools, grasslands, and pitch pine–scrub oak barrens. He estimates that between 300 and 1,200 Lenape lived on the island; faint spirals of smoke attest to their presence in several images.

Sanderson's hope has always been that once New Yorkers saw visions of Mannahatta dancing in their heads, they would think of their city in new ways. If they stepped into the past and appreciated the biological diversity of the island four hundred years ago, residents might want to preserve remaining ecosystems—of which there are many, and Sanderson includes photographs of examples taken in and around the city. Residents might embrace the creatures repopulating the city: the raccoons, red-tailed hawks, great horned owls, peregrine falcons, the occasional coyote. They might support ecological restoration or lobby for more parks and green space. "I guess my hope is that people fall so much in love with Mannahatta that they want to do those kinds of things. Not because they feel forced to, but because they want to," Sanderson says.

The final chapter of *Mannahatta* offers readers another thought experiment. Sanderson invites them to imagine the city they want in the future. His last line makes explicit his own desire: if the city makes thoughtful choices about transportation, energy consumption, food production, architecture, and green space, "Mannahatta will be back for all of us, a land as pleasant as one can tread upon, a city that all the people have created, connected by a thousand invisible cords, the fresh, green breast of a world that will thrive for another four hundred years, and then some." His closing illustration shows New York City 2409. Three greenbelts extend from Central Park, flowing down to the rivers; a stretch of Chelsea has gone to park. The Bronx, Queens, Brooklyn, and Staten Island have largely lost their human inhabitants and regained an agrarian past.[17]

Perhaps through the medium of information technology—through the quadrat of the computer screen or, if *Mannahatta* the app materializes, the phone or the handheld or the pad or the pod—a wired generation will imagine Manhattan wild. And if they do, they will be engaging an idea that was virtual before it played out in computers. "This is all about the virtual. The idea that nature is this trope we take to be our ultimate icon of the authentic or the real is profoundly human. Whether that idea finds its ideal in a *Mannahatta* computer program or a Hudson

River School painting by Frederick Church or a book like *Uncommon Ground*, all are human constructs," notes William Cronon. "*Mannahatta* is very much an expression of wilderness thinking in American culture, of a greater historical consciousness of changes in the landscape, and a sense of the human embedded in a natural matrix. *Mannahatta* is an invitation to mull on how a landscape can be both profoundly natural and human at the same time. Not that the two are not the same, but the way we tend to think of them is in opposition, as counterpoints."

The Garden in the Machine

Randel's survey of the Delaware & Raritan Canal was part of the national movement to lay down the country's infrastructure; it was to be, as Albert Gallatin intended, part of the necessary network of internal improvements. Canals cut through land, some of it farmed, some of it wild, bringing with them settlement, commerce, and industry. They were among the agents bringing the machine to the garden. Today the Delaware & Raritan Canal is a blue meander, a stretch of welcome water traveling across a suburban and industrialized landscape. It has become, as Howard Green of the New Jersey Historical Commission has noted, "the garden in the machine." Many early canals similarly have become significant recreational routes, the Erie and the Chesapeake & Delaware among them.

Many old railroad lines have become quiet bike paths, trails through the woods, refuges from highways, cars, and urban living. The High Line, which clattered and shed soot between factories and warehouses in lower Manhattan, transporting raw materials and produce, has become an aerial refuge, a long thin bridge of architectural and botanical beauty curving above the darting energy of the streets and sidewalks below. It too is a garden in the machine, one that explicitly honors its industrial history and its resurrection as a carefully cultivated landscape.

The High Line was built in the 1930s. By then many elevated lines wound above New York City, most of them carrying passengers. The possibility of traveling quickly above the streets, avoiding crowds and carriages, omnibuses and refuse, seemed a wild visionary scheme when presented to public audiences in the mid-nineteenth century. It reached those audiences alongside another wild visionary scheme: the elevator. Once adopted, the elevated railway permitted New York to expand and fill itself out, to ink in the pencil lines of the grid. The elevator permitted New York City to expand up and out of the grid, in a direction the street commissioners never described, doubtless never dreamed of. Disfavored natural hills and slopes found rebirth in the air as buildings brought people to new elevations. Manhattan remains an island of wild topographical variation—an island of layers.

Layers of time beguile us. In unfamiliar places, the present often crescendos while the past and future take a step back. There is no personal past to recall in a new place, no memory of an event that occurred at this corner or in that park. History may be explicit all around you—in the architecture, in stories of the place, in the landscape—but it is not your felt history. The present, however, necessarily grips you. You notice what is the same or different from where you live. You notice details along routes so you can find your way. You can see freshly because no habit or routine has yet dulled your senses, as it often does in familiar places.

If you return to a place you visited before—say it is Paris, but it might be Nairobi, Philadelphia, the house your grandmother lived in; it might be anywhere—you pass spots now quasi-familiar. You might remember that the last time you visited, you never imagined you would be back. You can see your younger self, your expectations, ideas, purposes, walking along a street unaware of your future self walking down that same street with different expectations, ideas, and purposes. Layers of experience can make that place feel nostalgic and magical, as though it has special significance. Time there feels less linear, more a looping, a backstitching. You wonder if you will be back this way again. In fanciful moments, you wonder if the place knew you would be back; if your future return was already set. And in this way you can feel the past, the present, and the future woven together. An experience of simultaneity arising from rare visits and memories evoked by place.

If you have lived a long time in one place, your personal history lays itself down more thickly, many strata in many places. The past is never far from the present. You see accumulations of your own experience, you feel flux, you see change. The layers of time we can experience in such a familiar place are for many of us an approximation, a hint, of what is experienced by ecologists, cosmologists, geologists, paleontologists, and others who explore the earth through time. The layers they see in a place can be centuries or eons thick. They can see the story of the earth's ever-shifting dynamo in the magnetic pull of a boulder. They can see the story of the earth's orbit and our solar system in the changing thickness and composition of ancient sediments making up the 222 million-year-old rock jutting out behind a New Jersey mall, where, just a stone's throw from Lowe's, fossil coelacanths can be hammered out of a hillside.[18]

VI

In Which Is Described "the Ingenuity of the New"

By 1852, New York City had extended up the island almost to 42nd Street. There, on the cutting edge of development, near the Croton Distributing Reservoir, now Bryant Park and the New York Public Library, the city sought to celebrate its technological prowess. Just a year before, Britain had held the first world's fair in the 990,000-square-foot Crystal Palace. Americans competing for prizes in the Great Exhibition of the Works of Industry of All Nations did better than anticipated. Mathew Brady, later famous for his Civil War photographs, won a medal for his daguerreotypes. Inventors won for an array of items, including a reaping machine, merino wool, and a fireproof safe. Americans were ready for the recognition. By midcentury, many felt eager to demonstrate that the New World could compete with the Old as a creative industrial power, as a force for innovation and invention. A group of prominent, forward-looking New Yorkers decided to capture this mood and assert equal footing with Britain. In March 1852 they formed the Crystal Palace Association and began planning the Exhibition of Industry of All Nations. In less than two years they had erected a 173,000-square-foot crystal palace to rival the one constructed in London. "There will be gathered here the choicest products

New York City's Crystal Palace. Courtesy of the collection of Mark D. Tomasko.

of Luxury of the Old World and the most Cunning Devices of the Ingenuity of the New," wrote Theodore Sedgwick, president of the association.[1]

As one commentator describes it, the exhibition represented the moment when American science and technology "modulated from a minor to a major key." Many innovators and inventors who presented at the American Crystal Palace in 1853 went on to fame, if not fortune, or their devices and ideas did. It was at the Exhibition of Industry of All Nations, later referred to as America's first world fair, that Elisha G. Otis debuted his safety elevator. Americans revealed particular skills in mechanization, measurement, and movement. They had mechanized many aspects of life, from object hoisting to bonnet pressing, knife polishing, brick molding, boot crimping, fruit paring and newspaper folding. They could survey, they could sound, and they could translate careful, accurate measurements into detailed maps, as the U.S. Coast Survey convincingly demonstrated at its display. And Americans could move people and things. Their work on canals and railroads and their need to travel great expanses had made movement and transportation areas of special attention and innovation.[2]

Transportation had become Randel's particular focus as well. He no longer counted himself among the surveyors, sounders, or cartographers. Instead he was one of three men showing designs for an elevated railroad for New York City. In Division A of the Crystal Palace—next to the Machine Arcade, where Benjamin Wright's son presented a model of a piston engine—Randel promoted with great energy a project he had been working on for just shy of a decade, his "Model of an elevated railway for Broadway, or other crowded thoroughfares." He had thrown all of his personal and family resources into creating a working iron model of his invention and trying to procure a city contract. He believed his elevated railroad would lift the quality of life in the city he had mapped. In his late fifties when he put forth the idea, Randel must also have realized he did not have much time left. The elevated railroad was Randel's final effort to establish himself as an American civil engineer and inventor of note.

Randel was not the first to have the idea for an elevated railroad in New York. Historian William Fullerton Reeves has documented six earlier proposals, including one made in 1832 by John Stevens, who had tested the first steam-run ferry on the Hudson River in 1804. Randel had surely traveled on Stevens's ferries to Hoboken when he worked on the island and visited Matilda and his family in New Jersey. As the population had surged in the early 1800s, the streets of New York City had become increasingly congested with carts and carriages, wagons and horse-drawn omnibuses; streets could be lethal to cross and were mucky and miasmic with horse manure and worse. Manhattan badly needed more, and improved, transportation. Stevens and others responded to this need early on, without success. Randel, who claimed he originally had the idea for an elevated rail line in 1829 but hadn't had a chance to perfect it until later, must have hoped his plan would be exciting and innovative enough for him to triumph where others had failed.

In 1846 Randel approached the Board of Aldermen for permission to build an elevated railroad on Broadway. (The Common Council had become the board in 1831.) He was joined in his petition by Gamaliel Gay, who a year later patented, with Charles J. Gilbert, a rubber vulcanizing apparatus. (The need for rubber that did not melt, crumble, or become rigid with use was growing with industrialization, and many people experimented with ways to, in essence, rubberize rubber. Charles Goodyear registered a patent for a vulcanizing process in 1843, and it was his process that eventually won out.) The aldermen seemed bewildered by Randel and Gay's proposal. First they assigned Randel and Gay's petition to the committee on arts, sciences, and schools. Then they assigned it to the street committee. Then they apparently threw up their hands and appointed a select committee of three men to review the proposal. When the special trio reported back the following month, they had much good to say: "The petitioners are in possession of a plan for constructing an elevated railroad that combines the very desirable advantages of strength, lightness of appearance and utility." But they were cautious. A decision of such magnitude could not be made quickly. Many people would have to view and vet such a massive transformation of the city's most prominent street. They instructed Randel and Gay to make models, write descriptions, and report back.[3]

The New Broadway over Broadway

Eighteen months later Randel, absent Gay, invited the aldermen to see his model on display at 413 Broadway. He commissioned the iron model in Philadelphia and then had it transported to New York. He also sent by special messenger invitations and tickets to every business owner and occupant along Broadway, "requesting them and their families to visit and examine those models and their appendages." Randel's proposal received much positive press over the coming years, and one of the earliest newspaper endorsements came from a New York correspondent for the *Daily National Intelligencer,* who reported that "new and magnificent plans are now under consideration" for Broadway. Randel's model for the "new Broadway over Broadway" was 30 to 40 feet long, according to the article, and 10 to 12 feet wide. (Randel's own reckoning put the model at one-tenth the size of the real thing, 60 feet in length and 2 tons in weight.) It had cost between $4,000 and $5,000 to make—roughly $113,000 to $141,000 today. "The perfect operation of the model is a pretty strong argument in favor of the plan," the correspondent concluded. "It has already been visited by many of our citizens, and, I believe, has generally made a favorable impression."[4]

Randel's design and models changed slightly over time, but the essential features remained consistent, as described in various newspaper accounts and in two pamphlets by Randel and by New York City's Mechanics Institute. The cars were to move at 6 miles an hour, pulled by "endless ropes" that were pulled in turn by stationary steam engines located in side streets. Cables pulled by fixed engines had been in use in England and Germany for decades; Randel would have been familiar with those designs. The cables would pull cars along four rail lines, running in a closed loop between Bowling Green and Union Place, 3 miles each way.

With those four tracks running in pairs, two running jointly uptown, two jointly downtown, Randel proposed a version of the local-and-express idea. An uptown passenger car would move continuously at 6 miles an hour on the express track. As that car entered a station, it would attach to a car alongside it, called a tender, which was waiting on the inner, local track, the one closest to the platform, and which was filled with passengers ready to go uptown. The two attached cars would move together along the platform, and for 90 seconds the doors between the two cars would open. Passengers wanting to get off would step into the tender and then later onto the platform; those wanting to go uptown would step from the tender into the uptown car. The express car would move the tender forward one station and drop it there. This design would permit the main cars to travel without stopping. "There must be as many of these tenders, with a *conductor to each*, as there are stopping places. These tenders, like the passenger cars, are to pass the whole circuit of the Railway; but this they do intermittingly, and not continuously, like the passenger cars," Randel wrote. "Each passenger car takes a *tender* from one landing to the next, and leaves it there to discharge its passengers and take in new ones by the time the next passenger car arrives opposite that station, when it is again carried forward one station, and so on."[5]

Randel's design was attentive to aesthetics, to the travel experience, to safety, and to economics. In one version, semi-transparent glass embedded in rubber would run in a walkway

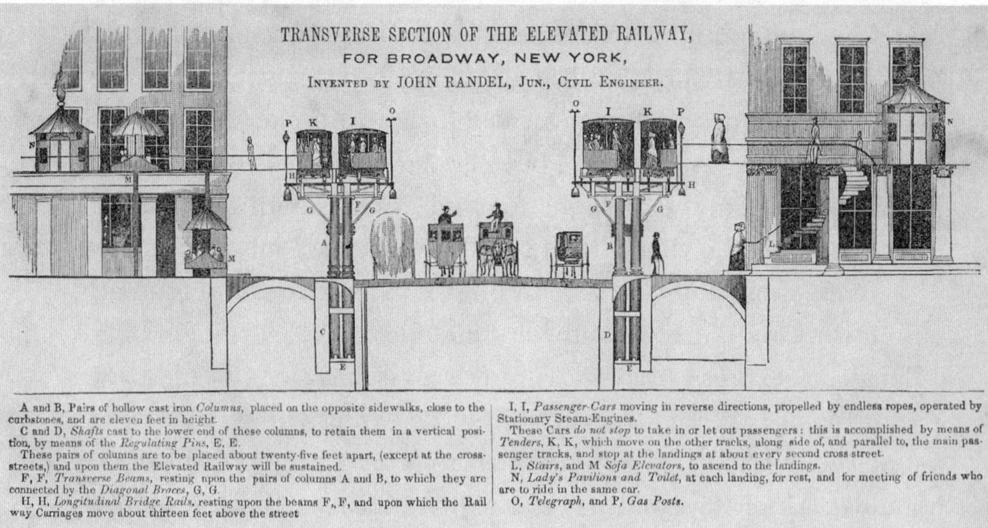

Elevated Railway by John Randel Jr. Courtesy of the New-York Historical Society.

above the sidewalks, alongside the elevated railroad, permitting "a pleasant and diffused" light to filter down to the street and doubling sidewalk capacity. (The tracks would run directly above the street, not the sidewalks.) In addition, Randel proposed painting the railway white and varnishing it, so it would "reflect *light* to the stores." He considered privacy in the buildings, suggesting that "elegant gauze wire window shades" could be installed to prevent passengers from peeking into homes. People traveling on his railroad could avail themselves of stairs or sofa elevators. (How those elevators worked is a mystery, as one twentieth-century commentator notes: Randel's sofa elevator is "possibly the first representation of a passenger elevator in this country and we are sorry that we haven't a clue as to how it was intended to work. The passenger elevator, surprisingly enough, is a most obscure subject.") Randel included a toilet and a ladies' pavilion at each landing; the pavilions were "to answer as places for rest, and for meeting of friends who are to ride in the same car." He described all the ways in which the elevated railroad was safe, both for those riding it and for those walking beneath it: "The usual accidents incident to the use of ordinary railroad cars, *cannot occur* on this Elevated Railway." And he was exacting about figures, including, as ever, detailed statistics on current transportation costs and revenue. Echoing his Erie Canal pamphlet in their precision and care, Randel's arguments were compelling, his research thorough. He had gathered information on how many omnibus licenses the city had granted, he had counted how many vehicles passed a certain point each day; he knew how many passengers traveled and when. He anticipated a return of between 8 and 23 percent on stockholders' investments. It is not clear what he would have netted, but if ridership met his expectations, annual gross revenues for the railway would have been more than $1 million, and presumably he would have been comfortably compensated.[6]

Randel even addressed concerns about health and filth. Every

New Yorker was preoccupied with such matters in the mid-nineteenth century. "In August 1853 alone, the city's contract scavenger reported clearing away 690 cows, 577 horses, 883 dogs, 111 cats, fourteen hogs, and six sheep—plus 1,303 tons of 'butchers offal' and sixty-two tons of refuse bones from slaughterhouses," Edwin G. Burrows and Mike Wallace describe in *Gotham*. That animal waste was joined by the daily contributions of more than 22,500 horses and the overflow from water closets. To address these concerns, Randel placed pipes for water, sewage, and gas under the street and provided receptacles for garbage. People could dump waste and offal into the lidded boxes. The small chambers in turn emptied into a vault: "The Disinfecting Vault (h) connects the Sewer Chambers on opposite sides of the Street . . . their Contents are to be removed and disinfected for Agricultural purposes: the Sale of which together with the Garbage will it is believed more than pay the interest on the cost of these improvements, besides promoting the Cleanliness and health of the City."[7]

Despite Randel's careful attention to every detail, despite his promising start with the aldermen, his first elevated railway proposal was rejected by the city. "The invention of Mr. Randell is one of great ingenuity, and no doubt can be applied to other cities with advantage, but should it come into use on Broadway, it would no doubt destroy the appearance of the street, as well as drive the citizens entirely from it." Another proposal coincident to Randel's—one Richard Varick DeWitt, Randel's former employer and friend, put forth—was also rejected. The aldermen decided to widen a few streets and increase the number of omnibuses instead—not the ideal solution for the "city of omnibuses," as New York had already been dubbed.[8]

Randel was not to be thwarted, however. He asked the Mechanics Institute to assess his invention and published their comments. The reviewers, including the president of the institute and the U.S. engineer at the Navy Yard in Brooklyn, applauded his plan, writing, "It appears to us that you have anticipated every possible contingency, and by ingenious mechanical arrangements and contrivances, have succeeded in securing all the utility and safety possible, without that experience which is only to be gained by practical operation. Should, however, new difficulties arise in the course of its construction and operation, we have no doubt, from the ingenuity already displayed, that you will be able to over come them." The chief engineer of the fire department gave his blessing as well. The railway would not interfere with the department's operations but might "be advantageously occupied by the firemen, and will to that height, aid in their successful operation upon fires."[9]

Randel altered his model slightly, presented another version to the Board of Aldermen, and exhibited it at an American Institute fair in Castle Garden. When those efforts still yielded no city contract, he devised a related plan that would not alter the appearance of Broadway but would improve city transportation in a different way. In June 1849, Randel proposed a belt road and railroad that would circle the island above the wharves. The 200-foot-wide "City Belt Railway and Depot" would be constructed along the North and East Rivers, from the Battery to Kingsbridge, filling in the river where needed to create more shoreline. A six-track elevated railway and promenade would run above the avenue, covered by a roof. "The City Belt Avenue, would, when completed, furnish a covered Railroad and Depot, about 28 miles in circuit, around the whole Island of Manhattan, or City and County of New-York, and make property on the East or North River as available as though it was located on the opposite river front," Randel described. It would be wonderful to ride, he argued, "in full view of our harbor and bay, interspersed with islands, surmounted by fortifications, together with the ships, steamboats, sloops and other river craft floating upon their waters, with the cities, villages, country seats and scenery on their opposite shores and in the distance." The aldermen liked this idea (as did city leaders in the twentieth century, who built an elevated highway along the Hudson River between the Battery and West 72nd Street). "Now this, it strikes ould be a most convenient and unique affair," a newspaper mented. But that project languished too.[10]

The same month Randel proposed the belt avenue railroad, he wrote to John M. Clayton, his former lawyer and the new secretary of state, about yet another scheme. Randel said he was busy at work on the design of a new city, Morrisania, in the Bronx. "I have made the following improvements in the Plan for this new city, when compared with the Plan of New York," he explained. "This City when built up, cannot fail to be the most beautiful and comfortable suburban City, you could desire to see. The Rail Road to Albany and New Haven pass through the middle of it." He seemed less sanguine about the elevated railroad at that moment. "I still expect my Elevated Railway will be adopted here at no very distant day, but it being a new and bold project, will require some time to form public opinion on the subject. I sent you drawings and descriptive pamphlets of it last fall by Dr. __ of Delaware which I hope you received."[11]

A Plan as Exactly Like His Own as One Full Moon Is to Another

In 1853, Randel again tried to captivate people with the idea of traveling above the street—perhaps hoping that public support would shift the aldermen's position or give them the courage to undertake a bold measure. At the Crystal Palace, in the heart of the first American celebration of technological innovation, he presented a new railway model, the "most eligible of all my plans." And he began to lay very public claim to his ideas, as he had decades earlier with William Bridges and Benjamin Wright, taking on three men who had used in their own designs elements that Randel contended were originally his.

These declarations of best invention began to appear in the

New York Tribune in the year of the exhibition. One came in early June from "a thoroughly practical man," John B. Wickersham, an iron manufacturer who was to exhibit drawings for his plan at the Crystal Palace just a month later. "The plans for constructing an Elevated Railroad in Broadway heretofore presented, have all possessed some grossly objectionable feature," he began his article. The next came a few weeks later from Elisha G. Otis. "It is in many respects like that of Mr. Wickersham, though in some particulars it seems to be an improvement on that plan," opined the editors. Then a Mr. J. R. Orton submitted his description. And on the same day Wickersham responded to Otis, saying that Otis had taken not only from him but from many others: "An Elevated Sidewalk and Railroad has been proposed or talked about, any number of times within the last twenty five years (as I have been informed *since* the publication of my first letter.) If Mr. Otis will procure the publication of Mr. John Randall Jr. on this same subject printed as early as 1848, he will find, on the 8th page of this interesting pamphlet, a plan as exactly like his own as one full moon is to another."[12]

Randel himself responded in late July. He would have responded earlier, he noted, "but have been waiting to see whether there would not be *some more claimants to those inventions of mine*." As always, he outlined in exacting detail the chronology of his invention, its mechanical attributes, and the praise given it by prominent men. The following day he sent a short note correcting some errors introduced to his letter by the *Tribune*, to wit: "In the first line of the second column, you have placed the figures 8 and 7 so close to each other, without a point or dash between them, as to read '*eighty-seven miles*' instead of its real distance of '*eight miles and seven-tenths*' in circuit from South Ferry to Crystal Palace and returning." A few weeks later, Randel found himself compelled to submit another letter as a fourth claimant had come forward, a Mr. Patrick O'Neil; "This plan is also a copy of my model," Randel wrote.[13]

Randel's fervor—and financial desperation—did not diminish.

He issued more particulars on cost and design in the *Tribune* in December 1853 and pushed again for the City Belt Avenue. He suggested that the state should legislate for the belt railway, as it had for the grid plan; he was hoping for a visionary political act like that of 1807 and a set of powerful patrons who could rise above petty objection. In July 1854 the *New York Tribune* ran Randel's advertisement for subscribers, urging them to come to the American Institute at 351 Broadway at 3:30 in the afternoon to invest in "the most eligible plan (of the seventeen) invented by Mr. RANDEL, for improving the city travel." Randel's elevated railroad was the clear favorite of the Mechanics Institute and, it appears, of the American Institute. His belt railroad idea had traction in Europe as well. "We call your attention to an American invention which has attracted the attention of London and Paris; I mean the elevated railway of John Randel, Jr., one of our most accurate engineers. His plan has the approval of the celebrated engineer of England, Brunel, who proposes to make the Belt railway all around London . . . And he is followed by most eminent French engineers, who propose a like splendid belt railway for Paris," reported the American Institute in 1856. "The idea of Mr. Randel seemed extravagant seven years ago, but not so more than several of the most magnificent applied discoveries of late."[14]

Randel had important endorsements and international acclaim. The city had desperate residents. "We cannot live here . . . and it is inconvenient to live across the arms of the sea on either hand. We want to live up town, or in the adjacent county of Westchester; and we want facilities for getting quickly, cheaply, comfortably from our homes to our work and back again," read an editorial in the *New York Herald-Tribune*. "Street Railroads and Omnibuses have their uses; but we have reached the end of them. They are wedged for hours at night and morning with men, women, boys and girls sitting, standing, and hanging on; it would be not be decent to carry live hogs thus, and hardly dead ones . . . Gentlemen of the Legislature! give us both the Underground and the Aerial Railways!"[15]

But Randel had no success. His inventions went unrealized. In 1868, three years after his death, the first elevated railroad opened on Ninth Avenue. Randel might have been relieved to know that the winning design wasn't from Orton, Wickersham, Otis, or O'Neil. Charles T. Harvey of Connecticut succeeded where the others had failed. Randel might have been especially pleased to read descriptions of Harvey's line as "shabby" and "frail-looking," as "the railway upon stilts, or the one-legged railway."[16]

RANDEL WAS FORCED to abandon his vision of travel in the air and of a clean walkable city bustling below. In the end he returned to surveying and cartography, the work that had made his reputation in the first place. He had spent everything he had—or didn't have—on the elevated railway, and spent more time battling lawsuits, which he inevitably lost. Randel was sued by two men who had worked on the elevated railroad models for him and were never paid. Both men said they had rushed to meet his deadline and expended great labor. His lawyer then sued him for nonpayment. He and Letitia put Randelia up for sale in 1854, but the entanglements with their neighbor Jacob C. Howard kept the land from yielding much money.

In 1863, unemployed and impoverished, Randel again approached the Board of Aldermen, proposing to make a set of 1,472 maps for the city on a scale of 25 feet to an inch—sixteen times greater than that of the farm maps—with many additional details from his field books. He estimated the project would take "7 years and 8 months of time" and cost $73,600. Two months later, the committee on arts and sciences reported back favorably. The members noted that Randel held the field books from his survey, they "being his own private property. They are accessible to no one else, are attainable from no other source, and unless secured to the public during the lifetime of Mr. Randel, will, in all probability be lost; in any event, their usefulness will be impaired, even though preserved after he passes away, as to the ordinary examiner, they would be nearly unintelligible." The committee unanimously recommended hiring Randel. New maps would prevent future expensive litigation, the committee said. And "by reproducing them, as now proposed, we reverse the ravages of time and restore more than forty years of valuable records."[17]

Other members of the board were not convinced. The recommendation was held over. Then it lost, three to nine. It was resurrected the following month and held over again. A few months after that, Mayor Charles Godfrey Gunther, whom Randel had likely approached independently, wrote to the board supporting Randel's petition; Randel had convinced several politicians of the value of his field notes (he had noted in a letter to the board that his knowledge may be "utterly lost, to the present and future generations," and that this was the city's "last opportunity.") To no avail. Perhaps because the petition did not pass and because it was clear that Randel's knowledge was profound and about to be lost, David T. Valentine, the long-time city clerk, asked Randel to write an essay about his grid work and had him draw some small maps locating the commissioners' office and Thomas Paine's residence. That ten-page essay, *City of New York, north of Canal street, in 1808 to 1821*, is the only narrative by Randel describing the grid survey.

Randel's final document appears to be a recollection he wrote for the editor of the *Historical Magazine*. "Sixty years ago, I was only a boy," he wrote, "yet I think I can comply with your request to tell you something of 'HARLEM, AS IT WAS FIFTY OR SIXTY YEARS AGO,' with some degree of accuracy. I will try to do so, with the understanding that if my friend and neighbor, Riker, shall hereafter find me in error, he and you will attribute it less to a desire on my part to misrepresent, than to a failure of my memory—for, to be candid, I am not what I once was." Randel died in Albany on August 2, 1865, of "brain inflammation." He was reportedly buried in the First Presbyterian Church of Orange cemetery, where Matilda was buried. But his gravestone is not among those still standing or still legible.[18]

BACKSIGHT AND FORESIGHT

Adversaries, like Benjamin Wright, and historians, like George Johnston, characterized Randel as fueled by a desire for recognition; Johnston called it "self aggrandizement," Wright called it "ever-lasting fame." Randel certainly trumpeted his achievements, from his exacting mapping of New York City and his inventions to his most eligible route for the Erie and the deep cut for the Chesapeake & Delaware Canal. He did the same with his elevated railroad, seeking to establish himself as the originator of the idea and certain elements of design. Through self-promotion—but also through the admiring accounts of many contemporaries—he did achieve a certain fame during his lifetime. In August 1858, when news that the transatlantic cable had been completed arrived in Saratoga Springs, Randel came to people's minds. "Suddenly I heard a great shout and huzza on the street; I looked out and heard a telegraphic dispatch had come from New York with the intelligence that the Atlantic Cable was completed and all right," wrote Eliza Potter in *A Hairdresser's Experience in High Life*. "Then came orders for a general jubilee; some began preparing speeches; some getting up different kinds of illuminations; some doing one thing, and some another, but none idle . . . After dark, the streets were full of bonfires, houses and stores were illuminated, fireworks of all kinds were set off; then came shouts for speakers; some called for Washington Irving; some for Mr. Cooley; some for Mr. Randel, of Philadelphia; some again for Gen. Cadwallader, and many others."[19]

Although his name is unfamiliar to most Americans today, Randel deserves to be more than a historical shadow. His careful surveys and fierce attention to detail gave rise to the grid map that has been, in large part, faithfully follwed by New York City planners. Whether you love the grid or not, it is clear that its uniform regularity never deadened the island; some even argue

John Randel 1854 Carte de Visite. Courtesy of the Maryland Historical Society.[21]

that its regularity sparks creative riffs. With his spectacular farm maps, Randel captured the island's nature just before the metropolis arose. His creative work on canals and railroads were an integral part of America's first transportation revolution. And his excellent data have given life to ecological imaginings today.[20]

There are more charitable ways to view Randel's motivations than did Wright or Johnson—equally valid ways. Randel was a visionary, in the modern and glorious sense of the word. He had a wonderful mind for detail and a love of precision. He loved math and he loved machines. He wanted his creations to advance and improve upon others and to work wonderfully so people would enjoy them, feel safe in them, thrive through them. He hoped his ideas would improve the world. He loved finding solutions for challenges, whether it was to walk a straight line through undulating terrain, design equipment to put Manhattan's hills on the level, excavate dirt from an embankment, or alleviate congestion in the city he had walked and imagined and sketched in his mind and on paper before anyone else had. As he walked his lines, Randel could see things behind him, he could see where he stood, and he could see well ahead.

EPILOGUE

Randel's Rock

On the morning of March 19, 2011, Randel's Central Park bolt has many unwitting visitors: tourists, a homeless man, dog walkers, and teenagers all amble across the rock, perhaps noting the bolt, most likely not. A jazz duo plays nearby, and bicycle cabbies describe the Mall and its literary statues—Robert Burns, Walter Scott, and William Shakespeare—to their passengers as they pedal by. The wind gusts across purple and lilac crocuses.

Shortly after noon, Lemuel Morrison arrives to spend some serious time with the bolt. He sets up his tripod and turns on the GPS receiver, which receives strong signals from six and sometimes eight of the nine satellites it can discern. It is Morrison's birthday. It is chilly; his jacket choice reveals unwarranted optimism about the arrival of warm sunlight and spring, and the rock gives no shelter from the wind. But Morrison waits until five in the afternoon before taking down his equipment and toting it away. He wants to make sure he records all the GPS data he can for several hours so the location of Randel's bolt can be made as precise as possible. For today Randel's long-forgotten bolt joins the nation's network of survey markers. It is now part of the National Spatial Reference System database. A bolt on a rock in a park on an island is connected to the satellites that travel above us in great arcs.

All those GPS data—all those satellites, all the corrections coming in from the nearby ground stations—will peg the bolt to its most exact location ever. Yet on another day, with perhaps all the same satellites and all the nearby reference stations transmitting corrections, Randel's bolt will have slightly different GPS coordinates. Fixed in our minds and in front of us, the bolt's precise location can never quite be pinned down. Every day the bolt's coordinates will shift, perhaps less than a centimeter. Morrison the surveyor, the purveyor of exactitude, loves this fact. "You will never know exactly, you will never get down to the truth," he says delightedly, lying back on the rock a few feet from Randel's bolt and looking up at the clear sky. "You will never know exactly where you are."

Central Park bolt. Illustration by Patricia J. Wynne.

ACKNOWLEDGMENTS

The genesis for this book was my husband, Tom Naiman, who told me about Eric Sanderson's *Mannahatta* project, which led first to a *New York Times* piece and then to a fascination with John Randel Jr. That fascination could never have been pursued without the support given me by Tom and our children, Auden and Julian. I am grateful for their love, good humor, generous and impish spirits, smarts, and sharp eyes. Just as the book went into proofs, Auden and Julian spotted a survey marker on Big Rock that I had never noticed.

When I began researching Randel, the bicentennial of the 1811 plan was well in the future and few people were intrigued by this obscure surveyor so key to New York City's form. But there were several, and they generously shared what they had discovered: Reuben Skye Rose-Redwood, J.R. Lemuel Morrison, Paul Cohen, and Caleb Smith. Over time, Rose-Redwood and Morrison became central to the story and have been trailed and peppered with questions for years now; this book would not exist without their intense participation. Many other people who appear in the book also repeatedly and patiently provided expertise: David Doyle, Hendrik Hartog, John Hessler, Jo Margaret Mano, David

Mladenoff, Eric Sanderson, and Deborah Warner. A special thank you to Jeffrey Lock for throwing himself wholeheartedly into interpreting Randel's instruments, and for his beautiful artwork.

Hunting for Randel took me to many archives, and I thank the following people for their help: Craig Carlson and staff at the Albany County Hall of Records; Tammis Groft and W. Douglas McCombs at the Albany Institute of History and Art; Jackie Penny and Thomas Knoles at the American Antiquarian Society; Nicolette Dobrowolski at Syracuse University; staff at the Cecil County Courthouse; Constance Cooper and colleagues at the Delaware Historical Society; Margaret Dunham and Bruce Haase at the Delaware Public Archives; Dan Ward at the Erie Canal Museum; Michael Dixon at the Historical Society of Cecil County; staff at the Historical Society of Pennsylvania, the Library of Congress, and the Maryland State Archives; James Green and Nicole Joniec at the Library Company of Philadelphia; Hector Rivera at the Manhattan borough president's office (who kindly pulled out the farm maps, time and again); James Amemasor at the New Jersey Historical Society; Bruce Abrams and Joseph Van Nostrand in New York City's Division of Old Records; Nicole Contaxis, Tammy Kiter, and Ted O'Reilly at the New-York Historical Society; Leonora Gidlund, Michael Lorenzini, and Kenneth Cobb at the New York City Municipal Archives; Thomas Lannon and Kate Cordes at the New York Public Library; William Gorman, Maria McCashion, Paul Mercer, and staff at the New York State Archives and the New York State Library; Richard Bennett, Alan Bauder, John Hernick, and Ralph Hill at the New York State Office of General Services; Karen Cooney and Thomas Burton at the Onondaga Historical Association; Aaron McWilliams, Jonathan Stayer, and staff at the Pennsylvania State Archives; and staff at the Rutgers University library. Special thanks to Mark and Nancy Tomasko for sharing their wonderful collection.

I am grateful to my hosts on several of those trips: Hannah Bloch, Abby Frankson, Johanna Keller and Charles Martin, Jean

Markowitz, and Bob Woodward and Elsa Walsh. Thanks to Lois Rockmaker. To Alisa Solomon and Marilyn Neimark for the car. And to all the Locks—Jill, Nicki, and Jeff.

I requested many at-distance searches and, again, I was kindly aided by many people: John McClintock at the Albany Academies; Amy Crumpton at the American Association for the Advancement of Science; Jovanka Ristic and Peter Lewis at the American Geographical Society; Charles Greifenstein at the American Philosophical Society; Susan McElrath at American University; Allen Tuten at the Central of Georgia Railway Historical Society; Joan Lieb at the Chenango County Historical Museum; Thomas Dixon at the Chesapeake and Ohio Historical Society; Stefan Bielinksi of the Colonial Albany Social History Project; Ana Guimaraes and Hilary Wong of Cornell University; Kathie Ludwig of the David Library of the American Revolution; David Wood at the First Presbyterian Church of Albany; Robert Reed of the First Presbyterian Church of Orange; Joyce Homan at the Genealogical Society of Pennsylvania; Melanie James at the General Society of Mechanics and Tradesmen; Rana Edgar at the Georgia Historical Society; Christopher Baer and Linda Gross at the Hagley Museum and Library; Donna Eschenbrenner at the History Center in Tompkins County; Georgina Compton at the UK Intellectual Property Office; Lori Chien at the Jervis Public Library; Christine Taft and Chris Olson at the Linda Hall Library; very special thanks to Bernice Coston at the Madison County Historical Society, who kept hunting for material and went out in the rain to photograph headstones; staff at the Library of Virginia; Eben Dennis, Jennifer Ferretti, and James Singewald at the Maryland Historical Society; Earlene Melious at Montgomery County's Department of History and Archives; T. Juliette Arai and Patrick Connelly at the National Archives; Stephen L. Gersztoff at the New York State legislative library; Sarah Wilcox at the New York State Historical Association Research Library; Richard Williams at the Oneida County Historical Society; Wayne Sparkman at the Presbyterian Church in America Historical Center; Lisa Jacobson and Leah Gass at the Presbyterian Historical Society; William Keeler at the Rochester Historical Society; Beverly Bradway at the Salem County Historical Society; Katherine Chansky at the Schenectady County Historical Society; Daniel Beams at the Schoharie County Historical Society; Courtney Esposito at the Smithsonian Institution archives; Mark Lambert at the Texas General Land Office; staff at the Texas State Library and Archives Commission; Kathleen Dow and Kate Hutchens at the University of Michigan; Nancy Shawcross at the University of Pennsylvania; and Regina Rush at the University of Virginia.

In addition to those cited in the text, several scholars and experts helped me: Tom Angotti, Hilary Ballon, Norma Basch, Markley Boyer, Dorothy Cmaylo, Elizabeth Covart, Joel Grossman, Nancy Hewitt, Eric Higgs, Sidney Horenstein, Matthew Jones, Bill Kane, Dennis Kent, Peggy Kidwell, Ronald Kline, Alfred Konefsky, Jessica Lautin, David Lowenthal, Charles Merguerian, Daniel Montello, Paul Olsen, Andrea Renner, Mary Ritzlin, David Smiley, David Stradling, Thomas Taber, and Joy Zedler. Lauren Mack worked on the bibliography. Connie Rosenblum and Vera Titunik of the *New York Times* shaped the *Mannahatta* article. Randel's relatives—Rand Weeks, Ann Weeks, and Lee Vosters—generously shared their family history. Kim Kastens suggested excellent sources and shared her work on geospatial thinking. Deirdre Baker's invitation to speak at the University of Toronto triggered my Big Rock recollection. Daniel Kevles provided encouragement and wisdom, recommended a camera that transformed my archive visits, and sparked an ongoing love for the history of science.

My colleagues at the Graduate School of Journalism at Columbia University contributed in innumerable ways. Samuel Freedman invited me to attend sessions of his renowned book course. Bill Grueskin and Laura Muha carved out time so I could research and write—and Laura shared in archival obsession. Helen

Benedict, June Cross, David Hajdu, LynNell Hancock, David Klatell, Dale Maharidge, Michael Schudson, Michael Shapiro, Duy Linh Tu, and Jonathan Weiner provided excellent advice and warm support. Sandro Stille shared his experience at every stage and was uncannily insightful and tough in the best possible way. Scott Osborn made sure I did not become repetitively stressed. Dian Zhou untangled technological problems and inspired me with her inventions. Sue Radmer shared her love of cartography. Cristina Ergunay helped me find archives, and Steve Toth curated dusty volumes from off-site. Richard John was buoyantly supportive and helpful, introducing me to critical sources and collections and to the Society for Historians of the Early American Republic. Deep thanks go to Alisa Solomon and Nicholas Lemann. Alisa for being my wise counsel, friend, and loyal Joe-and-roti comrade through this arduous process, which she went through in tandem as she wrote her book about *Fiddler on the Roof*. Nick for nurturing the idea, making recommendations about critical readings and pitfalls to avoid, and responding so thoughtfully to the first draft. His suggestions have been central to the shape this book has taken; without him, it would not have come to be.

The book had several additional outside readers and aids. John Hessler and Hendrik Hartog made many insightful and helpful comments. Tali Woodward's sharp editorial eye and critical reading improved the book, and her friendship sustained me. Lynn Berger dove into archives, gathered important material, and checked portions of the manuscript with intelligence, insight, and nuance; Lynn also introduced me to John Kouwenhoven's essays. Thank you very much John, Dirk, Tali, and Lynn.

Many friends listened, advised, and were supportive in myriad ways: Scott Blumenthal; Johnny Brandt; Christine Braunstein; Phoebe and Peter Eavis; Elly Eisenberg; Randi Epstein; Francesco Fiondella; Anne Georget; Stephen Hall; Ben, David, Emily, and Hilary Harris; Julie Hartenstein; Margaret Hoffman; Vikram Jayanti; Karl and Liz Katz; Alice Naude; Ulla Pors Nielsen; Kate Reuther;

Halley Harrisburg and Michael Rosenfeld; Sally Stapleton; Gwen Tarack; and Tom Watkins. Thanks to the lovely vines: Deborah Heiligman, Patricia Lakin, and Laurent Linn. And to Patricia J. Wynne, whose wonderful art is in the book.

Patricia also introduced me to Darrin Lunde and Bill Schutt, who kindly introduced me to my agent, Elaine Markson, and her colleague, Gary Johnson. Elaine shaped the proposal in important ways and found the absolutely right home for this book. My editor at Norton, Tom Mayer, has been a joy to work with: brilliant, wry, supportive, and challenging in equal measure. I am grateful for Tom's vision, raptor-like eye, creativity, and long-view equilibrium. The rest of the team at Norton has been stellar as well. Thank you Liz Duvall, Ryan Harrington, Nancy Palmquist, Rachel Salzman, Denise Scarfi, and Devon Zahn. Thank you very much Mark Melnick for the beautiful, absolutely right cover.

Bob Naiman, whose love and optimism have been unfailing throughout this often difficult process, remarked several times that he hadn't known before what a family project it is to write a book. He is, as ever, right. Profound thanks to Catherine McGeever, Claire McColgan, and Laurel Snyder, without whom life would not have run smoothly and Auden and Julian would not have remained so consistently happy; I am thankful for their love, warmth, and wisdom. Ralph Holloway shared his delight in, and expertise about, genealogy. Daisy Dwyer listened to my tribulations, invariably responding with an insightful or hilarious observation. Eric Holloway took on a heavy burden to give me the time and focus I needed. Anne Nesbet talked and walked with me from the initial idea on, while she wrote her beautiful book, *The Cabinet of Earths*. Caroline Sugg provided a clear, concise review of sine and cosine. And back to the point of beginning: another thank-you to my husband Tom, who is the only person to have read the manuscript as many times as my editor Tom did. I cannot imagine this book without the benefit of his sensitivity to language and probing questions.

Although neither of them can share in the finished book, Lee Naiman and Louise Holloway have a big presence in it. Lee gave so much love and unwavering support and asked such great questions. I miss her very much and know she would have felt joy to see the book completed. Louise, whom I also miss so much, set this all in motion. Her openness, love of books, and deep connection to the natural world made Big Rock magical.

NOTES

Abbreviations

NYC: Municipal Archives: New York City Municipal Archives
NYHS: New-York Historical Society
NYSA: New York State Archives

I. In Which Reuben Skye Rose-Redwood and J. R. Lemuel Morrison Set Out to Find the Imagined City

1 Heckscher, *Creating Central Park,* 36.
2. Moore, *A Plain Statement*, 49–50.
3. Augustyn and Cohen, *Manhattan in Maps,* 102.

II. In Which John Randel Jr. Affixes the City to the Island

1. JR Field Books, 61.2, courtesy of NYHS.
2. JR Field Books, 62.2, NYHS.
3. JR Field Books, 61.2, NYHS.
4. JR Field Books, 61.2, 63.3, NYHS.
5. Perry, *Scenes,* 6. Perry began his account: "There is, perhaps, no class of men who endure so many hardships and privations—whose fortitude and energy, whose intellectual powers, are taxed to a greater extent than the Government surveyor in his operations of the wild forests of the Southern and Western portions of the United States, in

paving the way for future wealth and aggrandizement of his country; and yet there is no class of men, of useful occupation, who receive a less share of consideration and sympathy."

6. JR Field Books, 59.1, NYHS.
7. Calhoun, *American Civil Engineer,* 26.
8. Details of Levi DeWitt's visit with John Randel Jr. in New York City come from an 1812 Randel notebook, courtesy of the collection of Mark D. Tomasko.
9. Hackett, *Rude Hand,* 63.
10. Charles Ellet to Mary Ellet, August 29, 1827, Ellet Papers, Special Collections Library, University of Michigan.
11. Hackett, *Rude Hand,* 67.
12. Worth, *Recollections of Albany,* 31; Hackett, *Rude Hand,* 74.
13. Carey, *Exhibit,* 7; JR Field Books, 64.1, NYHS
14. A scan of the title page of *The Elements of Mechanics* was graciously provided by Michael Dixon, historian at the Historical Society of Cecil County. It was published in 1832 by Mathew Carey (Carey & Lea) and dedicated to Clement Clarke Moore. Renwick was a member of the American Institute at the same time Randel was, and they were both on the Arts and Sciences Committee. For a report Randel wrote on a dry dock design for the Franklin Institute, see *The Franklin Journal and American Mechanics Magazine*, January 1827. Records at the American Association for the Advancement of Science show that Randel was elected in 1854 and was a member in 1855 and 1856. On Nathan Lanesford Foster, see Octavo vol. 63, Diary no. 87, Papers of Nathan Lanesford Foster 1804–1882, courtesy of the American Antiquarian Society.
15. Hackett, *Rude Hand,* 95–96.
16. *Albany Gazette*, April 14, 1808; Ogdena Fort (Mrs. Jerome Fort), notes, Madison County Historical Society; Sokoloff and Khan, "Democratization of Invention,"364. All conversions from nineteenth-century currency into 2011 values were made in mid-2012 at Measuring Worth: www.measuringworth.com/uscompare, using the purchasing power calculator.
17. Jesse Randel Jr. to Catherine Randel, January 2, 1834, and Catherine Randel to Abraham and Rebecca Randel, July 9, 1819, Madison County Historical Society.
18. Catherine Randel to Abraham and Rebecca Randel, July 9, 1819, Madison County Historical Society. Catherine chastised her sister-in-law on behalf of her own sister Rebecca: "Sister R writes you all have forgotten

that you have a sister that is a bereaved widow and with helpless babe among strangers, solitary and alone. I think it is cruel in the extreme to treat a sister in her situation with so much neglect."

19. Ibid.; Holden, *Sir William Herschel,* 114–15.
20. Thoreau worked as a surveyor from 1850 to 1862, apparently teaching himself through books such as Charles Davies's *Elements of Surveying and Navigation.* He also reproduced maps, seemingly in an effort to understand how the landscape and the people in it had changed. For more details, see Hessler, "From Ortelius to Champlain."
21. Linklater, *Measuring America,* 16–18.
22. Love, *Geodaesia,* n.p. See also Bedini, *Thinkers and Tinkers,* 41 and ch. 4. Bedini notes that although many textbooks on surveying came from England, "few of them . . . contained information useful in coping with the conditions of the New World, which were substantially different from those in England and Europe."
23. Snyder, *Mapping of New Jersey,* 68, 69. DeWitt and Erskine's New Jersey maps "are so accurate that they can be super-imposed over modern topographical maps of old roads still in use—and there are many—with only minor discrepancies," notes Snyder (74).
24. Simeon DeWitt to John Bogart, October 2, 1776, and July 25, 1779, in Bogart, *John Bogart Letters.* The letters also conveyed DeWitt's religious views; "You will find that all the enjoyment of this life except those which are purely rational and divine are Vanity and in the End prove a Vexation of spirit," he wrote to Bogart on August 1, 1781.
25. Bogart, *John Bogart Letters,* 41; Heidt, *Simeon DeWitt,* 8.
26. The original Military Tract was such poor land in the Adirondacks that veterans refused it, according to Jo Margaret Mano; the second allocation in the western part of the state was accordingly called the New Military Tract. Many of the townships there were given Roman and Greek names, such as Homer, Virgil, Manlius, and Cincinnatus, which DeWitt was credited with choosing. But he claimed no knowledge of the, as he put it, "obnoxious names"; Ristow, *American Maps and Mapmakers,* 78.
27. Munsell, *Annals of Albany,* 275.
28. See Mano, "Unmapping the Iroquois," 180. Military lands in upstate New York had been laid out in 7-square-mile parcels in 1781, before the national land ordinance, according to Johnson, *Order Upon the Land,* 42–44.
29. Johnson, *Order Upon the Land.* n.p. (preface); Linklater, *Measuring America,* 168.

30. Johnson, *Order Upon the Land,* 36; Johnson, "Rational and Ecological Aspects of the Quarter Section," 336. Johnson notes that Japan too has a history of a 36-based system. Certain forms, such as the circle, she writes, "have their roots in the human condition. The widely occurring use of the number six for subdivisions may, also, be linked to the primary directions in space 'natural' to humans, i.e., left to right, forward and backward, up or down"; see Ehrenberg, "Pattern and Process," 115.
31. Series A4016-77, vol. 12, folder 69, NYSA.
32. Mano, "Unmapping the Iroquois," 178.
33. Ibid., 186. Also see Marx, *Machine in the Garden,* 38: "Even in the sixteenth century the American countryside was the object of something like a calculated real estate promotion."
34. Journal entry of November 10, 1860, quoted in Hoy, "Thoreau as a Surveyor," 212, 217.
35. Forest clearing brought Americans "pleasurable excitement," wrote Basil Hall, a Scottish captain visiting the United States in the late 1820s. With a good ax the settler "sets merrily forward in his attack upon the wilderness . . . This passion for turning up new soils, and clearing wilderness, heretofore untouched by the hand of man, is said to increase with years. Under such constant changes of place, there can be very little individual regard felt or professed for particular spots. I might almost say, that as far as I could see or learn, there is nothing in any part of America similar to what we call local attachments. There is a strong love of country, it is true; but this is quite a different affair, as it seems to be entirely unconnected with any permanent fondness for one spot more than another." Hall, *Travels in North America*, 146.
36. Simeon DeWitt to John Randel, Jr., April 24, 1819, NYSA; Simeon DeWitt to John Randel, Jr., May 13, 1808, Rutgers University Archives; JR Field Books, 64.10, NYHS.
37. *National Intelligencer and Washington Advertiser*, July 12, 1805.
38. *New-York Evening Post,* April 8, 1814.
39. JR Field Books, 64.1, NYHS.
40. Series A4016-77, vol. 12, folder 69, NYSA; oaths of office, NYHS.
41. Series A0452, vol. 10, NYSA.
42. Ibid.
43. "Remonstrance against selling of grounds near government house," April 24, 1815, NYC Municipal Archives; Guernsey, *New York City,* 39.
44. Spanne, "The Greatest Grid."

45. *Minutes of the Common Council*, July 25, 1808.
46. Cajori, *Hassler*, 42; *Minutes of the Common Council*, February 16, 1807; *Minutes of the Common Council*, March 4, 1807.
47. Judd, *Untilled Garden*, 7. Judd is quoting Morris's *Notes on the United States of America* (Philadelphia, 1806).
48. *New-York Herald*, March 2, 1877; Stokes, *Iconography*.
49. JR Field Books, 61.2, NYHS.
50. Morris, DeWitt, and Rutherford, "Remarks of the Commissioners for Laying Out Streets and Roads in the City of New York"; Gouverneur Morris to Simeon DeWitt, October 2, 1807, Grantz Collection, Historical Society of Pennsylvania. The *Minutes of the Common Council* suggest that Morris was most active of the three commissions in directing the work, because his name appears most often in conjunction with it.
51. Randel, "City of New York, north of Canal street," 847.
52. JR Field Books, 68.1 (2), NYHS. In 1795 such transits were very rare. David Rittenhouse wrote to Simeon DeWitt that he knew of only two portable transit instruments in the states, one made by Andrew Ellicott and one by himself; Series A4016-77, vol. 18, folder 49, NYSA.
53. JR Field Books, 59.2, NYHS. Deborah J. Warner of the National Museum of American History and other experts of early surveying practices note that the names of instruments were quite fluid in Randel's day: one man's transit was another man's theodolite. An American surveyor's compass was a British circumferentor. Without the actual instrument in hand or an image thereof, it remains hard to know exactly what a surveyor was referring to. These descriptions of Randel's basic equipment are consistent with early nineteenth-century American techniques, but no images or descriptions tell us what kind of theodolite or telescope he used.
54. "Randel City Map Will Be Preserved," *New York Times*, May 21, 1993; JR Field Books, 68.1 (2), NYHS.
55. Valentine, *Manual of Old New York*, 841.
56. Ibid., 843.
57. Lamb, *History of the City of New York*, 571–72; Randel, "City of New York, north of Canal street,"848.
58. *John Mills v. John Randel Jr.*, 1810R-47, Division of Old Records, New York County Clerk's Office. "The standard language of trespass is very brute force and arms, very masculine," says Alfred S. Konefsky. "It is formulaic and almost poetic. And that was the standard issue language for centuries."
59. Hartog, *Public Property and Private Power*, 161. I am grateful to

Stephen Gersztoff at the New York State Legislature for locating a copy of "An act respecting Streets in the City of New-York," March 24, 1809, 32nd session, ch. 103.

60. *Albany Gazette*, September 22, 1820; Munsell, *Annals of Albany*, vol. 2, 150.
61. "Remarks of the Commissioners for Laying Out Streets and Roads in the City of New York," n.p. Morris's calculations of the number of monuments needed was incorrect. Presuming one monument per intersection, 155 streets and 12 avenues would yield 1,860 monuments.
62. *Minutes of the Common Council*, December 3, 1810.
63. Randel, "City of New York, north of Canal street," 839.
64. See John Randel Jr. to Mr. Thomas R. Mercein, December 3, 1813, NYC Municipal Archives. Calhoun, *American Civil Engineer,* 59, notes that "the importance of exact elevations and grades in this task furnished another distinction sometimes made between the engineer and the land surveyor. Ordinary surveying with its legal purposes had to meet precise standards only for horizontal measurements. Engineering had to seek also vertical precision, and the use of the leveling instrument was occasionally noted as a concrete act of marking the transition to engineering from some other occupation."
65. On true and magnetic north, I am indebted to Dennis Kent of Columbia University's Lamont-Doherty Earth Observatory, who explained the mechanism and checked for accuracy.
66. Quoted in Cajori, *Hassler,* 171–72.
67. Border disputes still arise over old surveys. For instance, Tennessee and Georgia periodically dispute their border, because an 1818 survey of the states' boundary does not accord with the official boundary. "Every 20 or 30 years, some bright legislator that doesn't know the background of this situation . . . will say, 'Oh, we're missing out on some of our territory,'" said Savannah-based historian Farris Cadle; Lee Shearer, "Tennessee-Georgia Border Dispute Derided," *Athens Banner-Herald*, March 3, 2008. See also Shaila Dewan, "Georgia Claims a Sliver of the Tennessee River," *New York Times*, February 22, 2008.
68. Bedini, *Thinkers and Tinkers*, 254; Simeon DeWitt, "Variation of the Magnetic Needle," *American Journal of Science and Art* (July 1829): 61; DeWitt, "On the Establishment of a MERIDIAN LINE," *Transactions of the Society for the Promotion of Useful Arts* 4 (1819): 26. For the recent data about the magnetic shift in the early nineteenth century, I again thank Dennis Kent.

69. John Randel Jr., "Annular Eclipse of the Sun of 16th June, 1806, and 26th May, 1854," *Weekly Herald*, May 20, 1854; Simeon DeWitt, "Observations on the Eclipse of 16 June, 1806," *Transactions of the American Philosophical Society* 6 (1809): 300; Simeon DeWitt, "Variation of the Magnetic Needle," *American Journal of Science and Arts* (July 1829): 62.
70. *Minutes of the Common Council*, September 5, 1808.
71. JR Field Books, 65.1, 65.4, NYHS.
72. John Randel Jr. to Mr. Thomas R. Mercein, December 3, 1813, NYC Municipal Archives; JR Field Books, 69 (1819) and 64.1, NYHS.
73. Randel cited the pamphlet in two places: on the 1821 map and in a draft letter to Peter Maverick, the engraver of the 1821 map, asking him to send copies of his drawings "that I may finish a written description of them"; JR Field Books, 64.1, NYHS.
74. *Explanatory Remarks and Estimates*; John Randel Jr. to Peter Mesier, October 6, 1813, NYC Archives; JR Field Books, 65.4, 69 (1819), NYHS; Randel, 1821 map of the City of New York, Library of Congress.
75. Randel, *Explanatory Remarks*. The angle of incline was vitally important to Randel. If he could not get a rod level, he would have to calculate triangles to get the true horizontal length. Several of Randel's notebooks have sine and cosine tables in them so he wouldn't have to do the calculations every time (he also used versed sines, or versines, a related formula not much used today), but the entire process still required time-consuming math. Referring to himself in the third person in a draft plea to the Common Council for more money, Randel described the labor. "If thereof he measured 2,000 feet per day after the lines were transited etc. he would have 67 triangles to calculate twice and add together to reduce that one day's work to horizontal measure and as many more additions to reduce that measure to a medium temperature thus giving for every mile of measure 170 triangles to be calculated twice and 3,800 figures to be added together, to do which would require nearly as many hours as it required to make the measurements and . . . one half of this work was required to be done each night (when he ought to have been at rest) to enable him to advance in measuring the next day . . . by this close application he became so much impaired as to make it necessary for him to stop work early in the fall." The Common Council wanted the work done as quickly as possible, he noted with a touch of melodrama, "least in cause of his death . . . the work might be delayed and perhaps not completed with the same care with which it was commenced." JR Field Books, 65.4, NYHS.

76. JR Field Books, 62.3, NYHS.
77. Roy, "An Account of the Measurement," 440, 58.
78. Hoare, *Quest for the True Figure of the Earth*, 157; Smith, *Introduction of Geodesy*, 23. Pumpkin and egg images are from Smith, 19.
79. JR Field Books, 61.2, 69 (1819), NYHS. Although Randel was conversant with spherical trigonometry, he did not use it consistently.
80. JR Field Books, 66.3, NYHS.
81. Randel surveyed north of 155th Street and set markers along Tenth Avenue. One bolt has survived on a steep, rocky, overgrown hillside.
82. JR Field Books, 69 (1816), 64.3, NYHS.
83. *Minutes of the Common Council*, March 10, 1817.
84. *Minutes of the Common Council*, March 10 and March 19, 1817.
85. JR Field Books, 63.2, 59.1, NYHS; Marino and Tiro, *Along the Hudson and Mohawk,* 1–2. The authors note that "quantifying spirit" is a translation of *esprit géometrique*, which was pervasive in eighteenth-century Europe: "Since numbers held out great promise for illuminating the workings of the natural and social orders, enlightened men and women took to measuring everything from air pressure to population. The United States would catch this fever later in the 1790s, in part through the influence of persons like [Count Paolo] Andreani."
86. JR Field Books, 63.2, NYHS.
87. JR Field Books, 65.3; 62.3, 59.4, NYHS.
88. JR Field Books, 62.2, 69 (1819), NYHS.
89. JR Field Books, 62.2, NYHS.
90. JR Field Books, 62.2, NYHS. In a Presbyterian church genealogy, Randel's birthday is given as December 4, 1787. But his own record contradicts that date.
91. JR Field Books, 60.4, 64.1, NYHS. On the possibility of William's trouble with drink, see the testimony of Nathan Boulden in the Chesapeake & Delaware Canal case; Harrington, *Reports of Cases Argued and Adjudged,* 296.
92. Ogdena Fort, notes, Madison County Historical Society.
93. Series A0452-79, vol. 13–17, NYSA.
94. JR Field Books, 64.1, NYHS.
95. JR Field Books, 60.4, 64.1, NYHS.
96. JR Field Books, 69 (1819); 64.1, NYHS.
97. JR Field Books, 64.1, NYHS.
98. Details about the portraits can be found in the Smithsonian Institution's Inventory of American Paintings and in Bolton and Cortelyou, *Ezra Ames of Albany.*
99. Ames expert Tammis K. Groft, deputy director and chief curator at the

Albany Institute of History and Art, confirms that these two portraits look like Ames's.

100. JR Field Books, 64.1, NYHS.
101. Ogdena Fort, notes, Madison County Historical Society; Ernenwein, *Verona*, 44.
102. Condict, *Her Book,* 36–37, 15, 40–41.
103. JR Field Books, 69 (1819), NYHS.
104. Ryan, *Cradle of the Middle Class,* 74; JR Field Books, 64.1, NYHS.
105. Klinghoffer and Elkis, "The Petticoat Electors," 169.
106. *Minutes of the Common Council*, April 8, 1810; April 22, 1811; May 13, 1811; November 23, 1812; JR, *New-York Evening Post*, April 8, 1814.
107. *New-York Evening Post,* March 21, 1814.
108. Ibid.
109. *New-York Evening Post*, March 24, 1814.
110. *New-York Evening Post*, April 8, 1814.
111. JR Field Books, 67.2, NYHS; *New-York Evening Post*, April 8, 1814.
112. *Minutes of the Common Council*, May 19, 1806; Augustyn and Cohen, *Manhattan in Maps*, 96–99.
113. *New York Commercial Advertiser*, May 7 and April 28, 1808; *Minutes of the Common Council,* March 7, 1808 (see also *New-York Evening Post*, July 1, 1808); *Minutes of the Common Council,* April 10, 1809, and March 12, 1810.
114. Augustyn and Cohen, *Manhattan in Maps,* 6.
115. Hall, *Travels in North America,* 101; Series A4016-77, vol 1, folder 100, NYSA.
116. Series A4016-77, vol. 1, folder 100, NYSA; JR Field Notes, 69 (1819), NYHS.
117. Series A4016-77, vol. 20, folder 20, NYSA.
118. Series A4016-77, NYSA; Harrington, *Reports of Cases Argued and Adjudged,* 285; JR Field Books, 65.3, 69 (1819), 64.1, 65.3, NYHS; *John Randel, Jr., appellant,* 104; JR Field Books, Onondaga Historical Association.
119. *New York Commercial Advertiser*, May 25, 1813.
120. Series A4016-77, vol. 15, folder 137, NYSA.
121. Series A4016-77, vol. 1, folder 100, NYSA. According to notes from Abraham Randel's granddaughter in the Madison County Historical Society, Oneida Castle was proposed as an alternate capital for New York, because it is in the center of the state, but it missed that designation by a vote of one in either 1815 or 1817. According to the *Oneida Daily Dispatch*, January 11, 2012, there were three attempts to make Oneida Castle the state capital.

122. Series A4016-77, vol. 1, folder 100, NYSA.
123. JR, "City of New York, north of Canal street," 848.
124. Moore, "A Plain Statement," 49.
125. Ibid., 43–44.
126. *Minutes of the Common Council*, November 9, 1812, and February 2, 1818.
127. JR Field Books, 69 (1819), NYHS.
128. Manuscript Group 1411, New Jersey Historical Society.
129. JR Field Books, 64.1, NYHS.
130. JR Field Books, 69 (1819), NYHS; *Minutes of the Common Council*, February 15, 1819. Matilda copied a few other maps for Randel, based on other surveyors' work. They are in the collection of the New-York Historical Society.
131. Stokes, *Iconography*, vol. 3, 564.
132. JR to Teunis Van Vechten, September 2, 1815, Albany County Hall of Records, SARA 1-232.
133. JR Field Books, 69 (1819), NYHS.
134. On the case, see *Randall against T. Van Vechten and Others* in Johnson, *Reports of Cases in the Supreme Court of Judicature.*
135. Series A4016-77, vol. 1, folder 100, NYSA.
136. JR Field Books, 69 (1819), NYHS.
137. Series A4016-77, vol. 1, folder 100, NYSA. According to John Hessler, senior cartographic librarian in the Geography and Map Division of the Library of Congress, insets are not unusual but scrolls very much are.
138. *New-York Evening Post,* March 5, 1821.
139. JR Field Books, 69 (1819); 64.1 (1820), NYHS.
140. JR Field Books, 60.4, NYHS.
141. Quoted in Carey, *Exhibit of the shocking oppression,* 7.

III. In Which Rose-Redwood Surveys the 1811 Grid and Morrison Surveys Today's

1. Hill and Waring, "Old Wells and Water-Courses," 370.
2. Information about various datums comes from "NGDV to NADV," Federal Emergency Management Agency, March 2007.
3. Zelenak, "The Xs and Ys of the Big Apple."
4. Rose-Redwood read Zelenak's article, which cited Koop. Koop, *Precise Leveling in New York City,* 71–72. According to Zelenak, Randel set the elevation at First Avenue and 27th Street because he thought that would be the center of the city as it grew.

5. Moore, *A Plain Statement*, 23.
6. Schuyler, *New Urban Landscape,* 23; Hartog, *Public Property and Private Power,* 159, 162.
7. Quoted in Jaye and Vatts, *Literature and the American Urban Experience,* 88.
8. Quoted in Rose-Redwood, "Rationalizing the Landscape," 104.
9. Quoted in Jackson and Dunbar, *Empire City,* 208; Schuyler, *New Urban Landscape,* 23.
10. Morris, DeWitt, and Rutherford, "Remarks of the Commissioners for Laying Out Streets and Roads in the City of New York"; Heckscher, *Creating Central Park,* 9; Isenberg, *Nature of Cities,* 95.
11. Rosenzweig and Blackmar, *The Park and The People,*135; Kostoff, *The City Shaped*, 74.
12. Simutis, "Frederick Law Olmsted," 281–82.
13. Quoted in Jackson and Dunbar, *Empire City,* 278–79.
14. Reps, *The Making of Urban America*, 299; Marcuse, "The Grid as City Plan," 287. It is interesting to note that very similar criticisms were made of the grids originating from the 1785 Land Ordinance. Landscape architect Horace Cleveland wrote in 1871 that "the monotonous character of their rectangular streets, which on level ground is simply tedious in its persistent uniformity, becomes actually hideous when it sets at defiance the plainest suggestion of natural topography and sacrifices every feature of natural beauty and every opportunity for picturesque effect in its blind adherence to geometrical lines." See Johnson, *Order Upon the Land*, 177.
15. Shanor, "New York's Paper Streets," 8–9; Rose-Redwood, "Rationalizing the Landscape," 62.
16. Morris, DeWitt, and Rutherford, "Remarks of the Commissioners for Laying Out Streets and Roads in the City of New York."
17. See Kostof, *The City Shaped,* 95: "The grid—or gridiron or checkerboard—is by far the commonest pattern for planned cities in history."
18. Hartog, *Public Property and Private Power,* 163, 165–66.
19. Cohen, *A Calculating People,* 149.
20. Rose-Redwood, "Rationalizing the Landscape," 74, 85.,
21. Adams, *Gouverneur Morris*, 282.
22. DeWitt, *Element of Perspective*, xix.
23. Rose-Redwood, "Rationalizing the Landscape," 85, 94. A similar argument was made by Elizabeth Blackmar in *Manhattan for Rent*. She notes that the grid suggested a revival of classical taste, which found beauty in symmetry and balance.

24. DeWitt, *Elements of Perspective*, 4, 25, 28.
25. Ibid.; Series A4016-77, vol. 1, folder 100, NYSA.
26. Rose-Redwood, "Rationalizing the Landscape," 94.
27. Rose-Redwood and Li, "From Island of Hills," 403.
28. This example is based on a helpful *NOVA* website called "GPS: The New Navigation."

IV. In Which Randel Keeps Seeking the Most Eligible Routes

1. Mathew Carey, Diary 1822–1826, November 2 and 22, 1825, Carey Collection, Library Company of Philadelphia, and the University of Pennsylvania Rare Book and Manuscript Library. In his November 22 entry, Carey states that all his effort has been "all to no purpose." But it was too early to conclude that.
2. Hall, *Travels in North America*, 153, 154–55.
3. Gallatin, "Report of the Secretary of the Treasury," 5, 8.
4. Ibid., 118.
5. Ibid., 85; Rubin, *Canal or Railroad?*, 9.
6. New Jersey Legislature, *Report of the Commissioners,* 12.
7. New York State, *Report of the Commissioners Appointed by Joint Resolutions,* 26–27, Albany Institute for History and Art; Larson, *Internal Improvement,* 76.
8. Hall, *Travels in North America*, 136.
9. Stilgoe, *Common Landscape of America,* 115. The figure for the tolls is for the Erie Canal and the feeder canals; Hood, *722 Miles*, 34.
10. JR Field Books, 64.1, NYHS; Dewitt quoted in Carey, *Exhibit of the shocking oppression,* 7.
11. Larson, *Internal Improvement,* 4.
12. JR Field Books, 64.1, NYHS.
13. JR Field Books, 69 (1819), NYHS.
14. JR Field Books, 1821–1836, Onondaga Historical Association.
15. JR Field Books, 69 (1819), NYHS.
16. New York State, "Report of the Commissioners," 16, Albany Institute of History and Art.
17. JR Field Books, 69 (1819), NYHS.
18. JR Field Books, 1821–1836, Onondaga Historical Association; JR, *Description of a Direct Route for the Erie Canal,* 5.
19. JR Field Books, 64.1, NYHS.
20. JR Field Books, 64.1, NYHS. In this instance, Randel publicly stated

several months later that Wright did not respond to his letters, so it seems reasonable to conclude that these two letters were sent in some form.

21. JR Field Books, Onondaga Historical Association; JR Field Books, 64.1, 64.10, NYHS.
22. "The Canal," *Albany Argus*, February 15, 1822.
23. Randel, *Description of a Direct Route for the Erie Canal*, 4.
24. Ibid., 8, 3; Calhoun, *American Civil Engineer,* 57.
25. Randel, *Description of a Direct Route for the Erie Canal,* 15.
26. JR Field Books, 69 (1819), NYHS.
27. JR Field Books, 69 (1819), NYHS.
28. Ibid. The troughs for loading boats come up frequently: in the field books in the summer of 1821, in the published Erie pamphlet, and during the Chesapeake & Delaware Canal work. Randel seemed well pleased with the idea, eager to see it tested.
29. Ibid.
30. JR Field Books, 64.1, NYHS.
31. John Randel Jr. to Bernard Peyton, December 26, 1822, and February 1, 1823, Library of Virginia. .
32. Gray, *National Waterway*. Gray's book is the major work of scholarship on this canal. It was only 14 miles long but made "possible continuous navigation along more than 600 miles of the Atlantic Coast without venturing to sea" (xvii).
33. Chesapeake & Delaware Canal directors to DeWitt Clinton, March 12, 1822, Chesapeake & Delaware Canal Papers, Delaware Historical Society; JR to J. Gilpin, March 17, 1823, C & D Canal Papers, Delaware Historical Society; JR Field Books, 64.1, NYHS.
34. JR Field Books, 64.1, NYHS.
35. Henry D. Gilpin to Joshua Gilpin, January 20, 1823, Gilpin Collection, Delaware Historical Society; Gray, *National Waterway*, 51.
36. Henry D. Gilpin to Joshua Gilpin, May 26 and 29, 1823, Gilpin Collection, Delaware Historical Society.
37. JR Field Books 64.1, NYHS.
38. Henry D. Gilpin to JR, August 20 and November 4, 1823, C & D Canal Papers, Delaware Historical Society.
39. John Randel Jr., "To Canal Contractors," *Times and Hartford Advertiser*, March 30, 1824; Gray, *National Waterway*, 52, 56.
40. Gray, *National Waterway,* 57; Harrington, *Reports of Cases Argued and Adjudged*, 292.
41. Benjamin Wright to John B. Jervis, July 9, 1824, Jervis Papers,

Jervis Public Library, Rome, NY. The identity of Old Father Putnam is unknown.

42. Harrington, *Reports of Cases Argued and Adjudged*, 297.
43. Gray, *National Waterway,* 58; Harrington, *Reports of Cases Argued and Adjudged,* 298.
44. Harrington, *Reports of Cases Argued and Adjudged,* 301.
45. Ibid., 293, 308–9. Cupping was a medical practice during which physicians would heat up the air inside a cup and then hold the cup face-down against the skin. As the air cooled, the idea went, it created a vacuum and pulled bad materials out of the skin and into the cup.
46. Henry D. Gilpin to Joshua Gilpin, October 1, 1825, Gilpin Collection, Delaware Historical Society.
47. Benjamin Wright to John B. Jervis, September 11 and November 20, 1825, Jervis Papers, Jervis Public Library, Rome, NY.
48. Quoted in Carey, *Exhibit of the shocking oppression and injustice,* 22–23.
49. Harrington, *Reports of Cases Argued and Adjudged,* 257.
50. Carey, "Protest," October 8, 1825, Library Company of Philadelphia.
51. Ibid.
52. See Koeppel, *Bond of Union,* 204–5: "We feel very much dissatisfied with the conduct of Mr. Wright."
53. A Stockholder, "Reply to Mr. Carey's Appeal," 2–3, Library Company of Pennsylvania; *New-York Spectator*, June 24, 1831.
54. Carey, *Exhibit of the shocking oppression and injustice*, 2, Library Company of Philadelphia. "When he arrived in Philadelphia, he ranked high in honour and reputation in his native state, New York, where he enjoyed the friendship and confidence of some of the most distinguished characters, who duly appreciated his energy and his talents—he was possessed of an independent fortune—in excellent health—a bright career of prosperity open before him . . . In a word, however ardent his ambition, there were few men with whom he could wish to change situations. Such *was* John Randel! What *is* he now? His reputation assailed by foul calumny, his professional talents depreciated; his fortunes and projects blasted . . ."—and on Carey goes.
55. Ibid., 6.
56. Carey, *Autobiography*, 4, 9, vi; Carey, "Essays on the Public Charities," 1.
57. Mathew Carey, diary, November 28, 1825, Library Company of Philadelphia, and the Rare Book and Manuscript Library, University of Pennsylvania; John Sergeant to Mathew Carey, 1825, Edward Carey Gardiner Collection, Historical Society of Pennsylvania.

58. Calhoun, *American Civil Engineer,* 98, 95.
59. Gray, *National Waterway*, 56.
60. *John Randel, Jr., appellant, vs. William Linn Brown,* 97.
61. Ibid., 87.
62. *Daily National Intelligencer*, June 3, 1831.
63. Harrington, *Reports of Cases Argued and Adjudged,* 43.
64. Howard, *Reports of Cases Argued*, 407; Harrington, *Reports of Cases Argued and Adjudged*, 178.
65. "Randel, Junior, vs. the Chesapeake and Delaware Canal Company," *National Gazette*, January 30, 1834.
66. *Baltimore Gazette and Daily Advertiser*, January 29, 1834; Comegys, *Memoir of John M. Clayton*, 16, 26–27.
67. Harrington, *Reports of Cases Argued and Adjudged,* 285, 289–90.
68. Bayard Collection, Box 61, folder 17, Delaware Historical Society.
69. Harrington, *Reports of Cases Argued and Adjudged*, 306–7.
70. *National Gazette*, January 30, 1834.
71. Carey, *Exhibit of the shocking oppression and injustice*, 7.
72. *John Randel, Jr., appellant, vs. William Linn Brown,* 111.
73. Ibid., 160.
74. Ibid., 100.
75. Rundell, *American Domestic Cookery*, 1, 8, American Antiquarian Society.
76. Quoted in Gifford, *Cecil County Maryland,* 140; Rundell, *American Domestic Cookery*, 111.
77. Charles Ellet to Mary Ellet, June 15 and November 4, 1827, Ellet Papers, Special Collections Library, University of Michigan. Randel apparently liked Ellet too. He wrote a letter of recommendation for him in 1828: "I found him to be a young gentleman of amiable manners, industrious habits; of strict integrity, sound discretion and good judgment; and he now has considerable experience in his profession: he is deserving of public and private confidence." See Lewis, *Charles Ellet, Jr.,* 13.
78. Larson, *Internal Improvements*, 225.
79. JR, "Railroad Experiments," *Baltimore Gazette and Daily Advertiser*, July 9, 1832; JR, report to the president and directors, July 4, 1832, New Castle & Frenchtown Railway Papers, Delaware Historical Society.
80. JR, report to the president and directors, July 4, 1832, New Castle & Frenchtown Railway Papers, Delaware Historical Society.
81. Ibid.
82. Ibid.
83. Lankton, "New Castle and Frenchtown Railroad," 8.
84. The first U.S. railroad to use an engine was the Mohawk and Hudson, just

five years before work began on the New Castle & Frenchtown. William F. Holmes's assessment of the New Castle & Frenchtown is more charitable than Lankton's. In "The New Castle and Frenchtown Turnpike and Railroad Company," Holmes contextualizes the company's challenges: "The building of a railroad—even though it was to be only a sixteen-and-a-half-mile track—presented a formidable problem in 1830 . . . But there were few railroad companies in the world to which the New Castle men could turn to for advice, and therefore they would have to be pioneers in helping to develop a new mode of transportation" (165–66).

85. Quoted in Hayman, *Rails Along the Chesapeake,* 7. Randel's great-nephew William R. Weeks was later piqued by Johnston's description. Writing in 1915, in response to an article about the railroad that cited Johnston, Weeks maintains that he has all Randel's papers—including a *Reminiscence of John Randel Jr.* written for his son, John Massey Randel—and that Randel's own account in that reminiscence indubitably reveals his skill, his "bull-dog tenacity," and "his usual thoroughness and progressive spirit." See Weeks, "Interesting Data."
86. Hayman, *Rails Along the Chesapeake,* 11.
87. *Report of the President and Directors to the Stockholders of the Ithaca and Owego Rail Road Company*, 104, American Antiquarian Society.
88. Weeks, "Interesting Data," 442.
89. *New-York Spectator*, March 3, 1834, citing the *Albany Daily Advertiser*.
90. *New-York Spectator*, July 15, 1831.
91. JR to William D. Lewis, February 7, 1832, New Castle & Frenchtown Railway Papers, Delaware Historical Society; *New-York Spectator*, July 22 and August 1, 1833.
92. Quoted in Burns, "History of the Ithaca and Owego Railroad," 39–40.
93. JR, "Report of the Engineer in Chief"; A. Merrill, "First Passenger Train in America," n.p.
94. A. Merrill, "First Passenger Railway in America," n.p.
95. J. Merrill, "History of the Development of the Early Railroad System," n.p.
96. Cmaylo et al., *Images of America: Verona*; notes, Madison County Historical Society.
97. JR, "Report of the Engineer in Chief"; Lee, *A History of Railroads in Tompkins County*, 7.
98. Dixon, "Central Railroad of Georgia," 54–55. Randel was approached by other railroad interests during the 1830s as well. Elkannah Watson, a prominent Albany businessman, wrote to Randel several times requesting his services on a short, proposed line between Port Kent and Keeseville;

E. Watson to J. Randel, April 30, 1832, April 6, 1833, and April 29, 1835, New York State Library, Manuscripts and Special Collections.

99. JR, "Chesapeake and Delaware Canal," *Niles Weekly Register*, February 28, 1835.

100. *John Randel, Jr., appellant, vs. William Linn Brown,* 87, 122, 66.

101. Ibid., 64–65.

102. JR, "Chesapeake and Delaware Canal," *Niles Weekly Register*, February 28, 1835.

103. "Randel and the Chesapeake and Delaware Canal," *Niles Weekly Register,* February 20, 1835.

104. *John Randel, Jr., appellant, vs. William Linn Brown,* 57.

105. Ibid., 18; Richard Varick DeWitt to John M. Clayton, March 14, 1843, Manuscript Division, Library of Congress. DeWitt wrote that he wished to close his accounts with Randel and was trying to track down a confidential agreement he had made with Randel "at a time when he had neither security to give, nor credit, nor friends to help him on in his suits against the Canal Company or Judge Wright."

106. Simeon DeWitt to Stephen Van Rensselaer, October 20, 1834, Rutgers University Archives.

107. New York State Assembly Report No. 209, April 2, 1849.

108. Orville L. Holley to JR, Series A4016-77, vol. 11, folders 33, 35, and 37, NYSA.

109. Ibid.

110. Although no record of their relationship has turned up, Randel had a great deal in common with his nephew. By 1839 Jesse had surveyed part of Texas with Richard S. Hunt, issuing a now famous and rare early map. "One of the seminal maps of the Republic period, this fascinating piece of cartography illustrated in bold details everything from original roads to early counties to the location of roaming wild horses and silver mines!" wrote a Texas state archivist. Jesse and Hunt published *The New Guide to Texas* (New York: J. Colton, 1839), the first guide for settlers of the new state, which "became standard for American settlers," according to *Going to Texas: Five Centuries of Texas Maps* (Fort Worth: Texas Christian University Press, 2007). In what sound like his uncle's words, Jesse (and Hunt) note that their map and guide are the only ones based on "accurate" surveys.

111. Mann, *Republic of Debtors*, 3.

112. Quoted in Boydston, *Home and Work,* 72.

113. George Washington to Simeon DeWitt, August 29, 1781, Rutgers University Archives.

114. Johnston, *History of Cecil County*, 391–92.
115. Ibid., 391; James McCauley, diary, May 19, 1857, Historical Society of Cecil County.
116. Johnston, *History of Cecil County*, 391.
117. JR to James Clayton, June 22, 1849, Manuscript Division, Library of Congress.
118. James McCauley, diary, 1858, Historical Society of Cecil County.
119. *Randal v. Howard*, U.S. Supreme Court Cases & Opinions, vol. 67, 1862. http://supreme.justia.com/cases/federal/us/67/585/case.html.

V. In Which *Mannahatta* Lifts Off

1. Nash, *Wilderness and the American Mind*, 43. "Wilderness was the basic ingredient of American culture. From the raw materials of the physical wilderness, Americans built a civilization. With the idea of wilderness they sought to give their civilization identity and meaning" (xi).
2. For more on the country's early attitudes toward nature, see Judd, *Untilled Garden*.
3. Marsh, *Man and Nature,* 36.
4. Cronon, *Changes in the Land,* 122.
5. Marsh, *Man and Nature,* 35.
6. An account of Olmsted hearing Marsh lecture can be found in Roper, *FLO*, 11. But Mark Stoll of Texas Tech University, who is looking into this claim, has not been able to verify it. Spirn, "The Authority of Nature," 104; quoted in Brookline GreenSpace Alliance, Fall 2009, 6. www.brooklinegreenspace.org/index.html.
7. Augustyn and Cohen, *Manhattan in Maps,* 84.
8. The Mohawk name for the island did honor the many wetlands: *gänóno* means "reeds" or "place of reeds." See Shorto, *Island,* 42.
9. Kassulke, "See Wisconsin Through the Eyes of 19th-Century Surveyors." Mullett's and many other surveyors' entries can be viewed on Wisconsin's Board of Commissioners of Public Lands website, http://digicoll.library.wisc.edu/cgi-bin/SurveyNotes/SurveyNotes-idx?type=div&byte=78113&isize=L&twp=T010NR007E.
10. Viele, *Topography and Hydrology of New York*, 4, 12.
11. For a reproduction of Viele's water map and for Viele's use of Randel's data, see Augustyn and Cohen, *Manhattan in Maps,* 136–39. On flooding, see Steven Kurutz, "When There Was Water, Water Everywhere," *New York Times*, June 11, 2006.
12. Robert Juet, *Purchas His Pilgrimes* (1624), 591–92, http://documents.nytimes.com/robert-juet-s-journal-of-hudson-s-1609-voyage.

13. Burrows and Wallace, *Gotham,* 3–4.
14. Wilson, *Biophilia*, 22.
15. The quadrat appears to have first been used as an ecological tool by H. Hoffman in 1879 in Germany, but at a very large scale: each quadrat was 21.4 square kilometers; Tobey, *Saving the Prairies*, 51.
16. Others have felt Sanderson's impulse. In 1978 *The New Yorker* carried a piece about artist Alan Sonfist, who restored a corner of original Manhattan. Sonfist had grown up near the Bronx River and, as the article describes, loved being in the woods there: "He felt that the trees were as alive as he was, he wondered whether the rocks had a language of their own, and he understood that human beings don't have to use a forest, or ever do anything in a forest to enjoy themselves: they only have to be there, and look and listen." Sonfist reconstructed three slices of New York's pre-European landscape—a meadow, a young forest, and a mature forest—using native species, whose names he gleaned from historical research. Today his Time Landscape endures at the intersection of LaGuardia Place and West Houston Street. In the early 1990s archaeologist and GIS expert Joel W. Grossman began designing virtual reconstructions of Manhattan's seventeenth-century vegetation and animal life. His three-dimensional images—incorporating information from paleobotanical records, excavations of Dutch gardens in lower Manhattan, and his own topographical ground-truthing—portray the island green and hilly with trees, waterways and soaring birds; see www.geospatialarchaeology.com/3dnyc.html.
17. *Mannahatta* has continued to evolve since 2009. In 2010 *Mannahatta* was incorporated into *Welikia* (the Lenape word for "my good home"), which will describe the historical ecology of the other four boroughs as well. Doing this is both more and less complicated, Sanderson notes. The GIS model already exists, so he doesn't have to build one; on the other hand, there is no British Headquarters Map or Randel information for the other boroughs, no single surveyor or cartographer with the necessary topographical and hydrological data. In 2011 the Rockefeller Foundation funded Sanderson to build an interactive 2409 site, which will enable New Yorkers to visualize how lifestyle, urban planning, restoration ecology, and other elements can alter how much carbon they release. And Colleen Macklin, her students at Parsons New School for Design, and Sanderson just created a prototype of *Mannahatta: The Game* for iPhone. Players can rise through the ranks as "Eco-Masters" if they travel the island with an ecologically savvy eye, making ecologically savvy choices.

18. I am grateful to paleontologist Paul E. Olsen of Columbia University for his wonderful work and lectures on this topic, and for sharing this surprising fossil site in New Jersey with the M.A. science journalism seminar.

VI. In Which Is Described "the Ingenuity of the New"

1. New York City Crystal Palace Records, Box 28a, folder 3, NYHS; Post, "Reflections of American Science and Technology," 338.
2. Post, "Reflections of American Science and Technology," 338.
3. *Proceedings of the Board of Aldermen* (Bryant & Co.), vol. 30, February 19 and March 23, 1846.
4. "Elevated Railway in Broadway," *Daily National Intelligencer*, November 4, 1847. Another newspaper article gives different dimensions and cost; according to the *Cleveland Herald* on August 2, 1847, Randel's model was 31 feet long and cost $3,000. And the *American Railroad Journal and General Advertiser* of July 24, 1847, reported that Randel intended to take his model to London "where a patent has been secured." Science historian and patent expert Daniel Kevles of Yale University notes that American inventors would often seek to protect an invention by patenting it in England if they thought there were already U.S. patents on the design. According to the U.K.'s intellectual property office, there is no record of such a Randel patent. The U.S. Patent and Trademark Office does not appear to have his patent either.
5. JR, *Elevated Railway*, 10.
6. JR, *Explanatory Remarks and Estimates*, 6–8, 12; Peterson, "The Sofa Elevator," 31; JR, *Elevated Railway*, 13.
7. Burrows and Wallace, *History of New York City,* 786–87; "View of Broadway in the city of New York," lithograph by Robert J. Raynor, I. N. Phelps Stokes Collection of American Historical Prints, New York Public Library. The elevated railway description provides the most complete of just three records of Randel's involvement with public works projects. In 1836—as noted in Chapter IV—he had presented a report to the Baltimore city government on water supply, suggesting where to best to run pipes into the city and from which source. His other, and final, engagement with utilities may have arisen directly out of the elevated railway proposal and Randel's knowledge that New York City's government was trying to address the sewage disaster by building new pipelines and repairing and redesigning older, ineffective ones. In February 1854 he approached the Board of Aldermen with

an idea for sewer, water, and gas pipe chambers running under the street. His petition seems to have been picked up or commissioned by professor Lewis A. Sayre, a resident physician for New York City and a founder of the American Medical Association. Sayre suggested a system of sewer, water, and gas pipelines that would run under the streets and sidewalks and would be accessible via trapdoors; Randel, "one of the ablest civil engineers the country has ever produced," drew up or designed Sayre's plans. The plans sound remarkably similar to those Randel depicted in his 1848 elevated railway proposal; they are described in "The Sanitary Topography of New York City" in *The Catholic World: A Monthly Magazine and General Literature and Science* 10, no. 59 (December 1869).

8. *Proceedings of the Board of Aldermen*, vol. 34, December 6, 1847.
9. JR, *Elevated Railway,* 18–19.
10. *New York Tribune*, December 28, 1853; "City Items," *New York Tribune*, June 8, 1849; *Proceedings of the Board of Aldermen,* vol. 37, June 13, 1849.
11. JR to John M. Clayton, June 22, 1849, Manuscript Division, Library of Congress. Morrisania, he described, would be more luxurious and spacious than New York City: "The *avenues* on *New York island* are all *100* feet wide & the cross streets at right angles thereto 60 feet wide . . . While in *Morrisania City*, the avenues are laid out *200 feet wide,* with *Parks* for trees and tasteful shrubbery, 80 feet in width in the middle, leaving a street 60 feet wide on each side of it."
12. "A New Broadway Over Broadway," *New York Tribune*, June 11, 1853; "An Elevated Railway for Broadway," *New York Tribune*, June 21, 1853; "Elevated Broadway Railroad," New York Tribune, June 28, 1853.
13. "An Elevated Railway for Broadway," *New York Tribune*, July 25, 1853; "City Railroads—Broken Car Wheels," *New York Tribune*, September 7, 1853.
14. "Relief for Broadway," *New York Tribune*, December 28, 1853; "Randel's Elevated Broadway and Promenade: Notice," *New York Daily Times,* July 17, 1854; *Transactions of the American Institute of the City of New York for the Year 1855*, 109; "Science, Mechanics, and the Fine Arts," *Daily National Intelligencer*, October 16, 1855.
15. Quoted in Hornung, *The Way It Was,* 118. The entire editorial or unsigned letter appeared in the *New York Herald-Tribune* on February 2, 1866.
16. Hood, *722 Miles,* 49. The elevated lines contributed to the expansion of the city, allowing more commuters to work and live in different places,

and were popular with those who rode them. But to many others they were an unsightly, ungainly solution to transportation problems. "The elevated structures darkened and obstructed the streets, and passing trains at the level of second stories were extremely noisy. Locomotives' stacks issued smoke and cinders, while passing trains dripped lubricating oil down onto the street; altogether the lines were a remarkable intrusion on urban neighborhoods, built for the convenience of those who passed through at the expense of those who lived there." Stradling, *Nature of New York,* 112.

17. *Proceedings of the Board of Aldermen*, vol. 92, October 5, October 12, October 26, and November 12, 1863; vol. 94, May 30, 1864. See also City Clerk papers from 1863, New York Municipal Archives.
18. Randel, "Old New York Revived," 27–30. The date given for Randel's contribution is June 1867, but that must be a mistake, because the August 19, 1865, *New York Herald-Tribune* reports Randel's death as occurring in Albany on August 2. An Albany database gives August 1. Brain inflammation may have been encephalitis or typhus. See Munsell, *Annals of Albany*, regarding the Orange, N.J., burial.
19. Potter, *Hairdresser's Experience,* 91. The identification of John Randel Jr. comes in the recent annotated version; see Santamarina, *Hairdresser's Experience*, 205.
20. "The outcome is a dazzlingly precise creative unity," writes John Kouwenhoven of the Manhattan skyline; *Beer Can by the Highway,* 52.
21. The similarities between this photograph and Randel's portrait by Ezra Ames are striking, and the name, date, and place all correspond. But the black-and-white scan of the portrait is unclear in some details (such as eye color). And there is always a chance this photograph is of another John Randel in 1854.

BIBLIOGRAPHY

Collections

Albany County Hall of Records, New York
Albany Institute of History and Art, New York
American Antiquarian Society, Worcester, Massachusetts
American University Archives/Special Collections, Washington, D.C.
Bureau of Land Management, New York State Office of General Services, Albany
Cecil County Courthouse, Elkton, Maryland
Central of Georgia Railway Historical Society
David Library of the American Revolution, Washington Crossing, Pennsylvania
Delaware Historical Society, Wilmington
Delaware Public Archives, Dover
Erie Canal Museum, Syracuse, New York
E. S. Bird Library, University of Syracuse, New York
First Presbyterian Church, Albany, New York
Georgia Historical Society, Savannah
Hagley Museum and Library, Wilmington, Delaware
History Center in Tompkins County, Ithaca, New York
Historical Society of Cecil County, Elkton, Maryland
Historical Society of Pennsylvania, Philadelphia
Jervis Public Library, Rome, New York
Library Company of Philadelphia

Library of Congress
Library of Virginia, Richmond
Linda Hall Library of Science, Engineering, and Technology, Kansas City, Missouri
Madison County Historical Society, Oneida, New York
Manhattan borough president's office, New York
Maryland Historical Society, Baltimore
Maryland State Archives, Annapolis
National Museum of American History, Washington, D.C.
New Jersey Historical Society, Newark
New York City Division of Old Records
New York City Municipal Archives
New-York Historical Society, New York City
New York Public Library
New York State Archives, Albany
New York State Library, Albany
Onondaga Historical Association, Syracuse, New York
Pennsylvania State Archives, Harrisburg
Rutgers University, archives and special collections, New Brunswick
Salem County Historical Society, Salem, New Jersey
Texas General Land Office, Austin
Texas State Library and Archives Commission, Austin
University of Michigan, special collections library, Ann Arbor
University of Pennsylvania, Rare Book and Manuscript Library, Philadelphia
University of Virginia, Albert and Shirley Small Special Collections Library, Charlottesville
U.S. National Archives and Records Administration, Washington, D.C.

Adams, William Howard. *Gouverneur Morris: An Independent Life*. New Haven: Yale University Press, 2003.

Adler, Phoebe, Tom Howells, and Duncan McCorquodale, eds. *Mapping New York*. London: Black Dog, 2009.

Allen, David Y. "How Simeon DeWitt Mapped New York State." 2008. www.newyorkmapsociety.org/FEATURES/ALLEN2.HTM.

Allen, Thomas G. *Memoir of the Rev. Benjamin Allen, Late Rector of St. Paul's Church*. Philadelphia: Latimer, 1832.

Andrews, William J. H. "A Chronicle of Timekeeping." *Scientific American*, January 2012, 50–57.

Augustyn, Robert T., and Paul E. Cohen. *Manhattan in Maps: 1556–1995.* New York: Rizzoli, 1997.

Baer, Christopher T. *Canals and Railroads of the Mid-Atlantic States, 1800–1860*. Wilmington, DE: Regional Economic History Research Center/Eleutherian Mills-Hagley Foundation, 1981.

———. *A General Chronology of the Pennsylvania Rail Road Company, Its Predecessors and Successors and Its Historical Context*. Pennsylvania Technical and Historical Society. 2004–2011. www.prrths.com/Hagley/PRR_hagley_intro.htm.

Balfour, Alan. *World Cities New York*. New York: Wiley, 2001.

Ballon, Hilary, ed. *The Greatest Grid: The Master Plan of Manhattan, 1811–2011*. New York: Columbia University Press, 2012.

Ballon, Hilary, and Kenneth T. Jackson, eds. *Robert Moses and the Modern City: The Transformation of New York*. New York: Norton, 2008.

Barrett, Walter. *The Old Merchants of New York City.* New York: Greenwood, 1968.

Bean, W. T., and E. Sanderson. "Using a Spatially Explicit Ecological Model to Test Scenarios of Fire Use by Native Americans: An Example from the Harlem Plains, New York, NY." *Ecological Modelling* 211 (2008): 301–8.

Beans, E. W. *A Manual for Practical Surveyors, Containing Methods Indispensably Necessary for Actual Field Operations*. Philadelphia: J. W. Moore, 1854.

Bedini, Silvio A. *Early American Scientific Instruments and Their Makers.* Washington, DC: Smithsonian Institution Press, 1964.

———. *Thinkers and Tinkers: Early American Men of Science.* New York: Scribner's, 1975.

Benjamin, Marcus. *A Historical Sketch of Madison Square*. New York: Meriden Britannica, 1894.

Bennett, J. A. *The Divided Circle: A History of Instruments for Astronomy, Navigation and Surveying.* Oxford: Phaidon-Christie's, 1987.

Berkeley, Edmund, and Dorothy Smith Berkeley. *George William Featherstonhaugh: The First U.S. Government Geologist*. Tuscaloosa: University of Alabama Press, 1988.

Berman, John S. *Portraits of America: Central Park. Museum of the City of New York*. New York: Barnes & Noble, 2003.

Bernstein, Peter L. *Wedding of the Waters: The Erie Canal and the Making of a Great Nation.* New York: Norton, 2005.

Berthon, Simon, and Andrew Robinson. *The Shape of the World: The Mapping and Discovery of the Earth.* New York: Rand-McNally, 1991.

Black, George Ashton. *The History of Municipal Ownership of Land on Manhattan Island to the Beginning of Sales by the Commissioners of the Sinking Fund in 1844*. New York: AMS, 1967.

Blackmar, Elizabeth. *Manhattan for Rent: 1785–1850*. Ithaca: Cornell University Press, 1989.

Blayney, J. McClusky. *History of the First Presbyterian Church of Albany, N.Y.: Lists of Its Officers, and a Complete Catalogue of Its Members from Its Organization.* Albany: Jenkins & Johnston, 1877.

Bloodgood, Simeon Dewitt. *An Englishman's Sketch-Book; Or, Letters from New-York*. New York: G. and C. Carvill, 1828.

Boelhower, William. "Inventing America: A Model of Cartographic Semiosis." *Word & Image* 4, no. 2 (April–June 1988): 475–97.

Bogart, John. *The John Bogart Letters: Forty-two Letters Written to John Bogart of Queen's College, Now Rutgers College, and Five Letters Written by Him, 1776–1782, with Notes.* New Brunswick: Rutgers University Press, 1914.

Bolton, Reginald Pelham. *Indian Paths in the Great Metropolis*. New York: Museum of the American Indian/Heye Foundation, 1922.

Bolton, Theodore, and Irwin F. Cortelyou. *Ezra Ames of Albany: Portrait Painter, Craftsman, Royal Arch Mason, Banker, 1768–1836*. New York: New-York Historical Society, 1955.

Boyd, Lois A., and R. Douglas Brackenridge. *Presbyterian Women in America: Two Centuries of a Quest for Status.* Philadelphia: Presbyterian Historical Association/Greenwood, 1996.

Boydston, Jeanne. *Home and Work: Household, Wages and the Ideology of Labor in the Early Republic*. Oxford: Oxford University Press, 1990.

Breed, Charles Blaney, and George Leonard Hosmer. *The Principles and Practice of Surveying.* New York: Wiley, 1906.

Bridges, William. *Map of the city of New-York And island of Manhattan with Explanatory Remarks and References*. New York: T. & J. Swords, 1811.

Brown, Henry Collins. *Fifth Avenue Old and New: 1824–1924*. New York: Fifth Avenue Association, 1924.

Buell, Lawrence. *The Environmental Imagination: Thoreau, Nature Writing, and the Formation of American Culture*. Cambridge, MA: Belknap/ Harvard University Press, 1995.

———. *Writing for an Endangered World: Literature, Culture, and Environment in the U.S. and Beyond*. Cambridge, MA: Belknap/ Harvard University Press, 2003.

Burns, Michael John. "The History of the Ithaca and Owego Railroad." Master's thesis, University of Delaware, 1966.

Burrows, Edwin G. *Forgotten Patriots: The Untold Story of American Prisoners During the Revolutionary War*. New York: Basic, 2008.

Burrows, Edwin G., and Mike Wallace. *Gotham: A History of New York City to 1898*. New York: Oxford University Press, 1999.

Buttenwieser, Ann L. *Manhattan Water-Bound: Manhattan's Waterfront from the Seventeenth Century to the Present*. New York: Syracuse University Press, 1999.

Cajori, Florian. *The Chequered Career of Ferdinand Rudolph Hassler, First Superintendant of the United States Coast Survey: A Chapter in the History of Science in America*. Boston: Christopher, 1929.

Calhoun, Daniel Hovey. *The American Civil Engineer: Origins and Conflict*. Cambridge, MA: Technology Press, 1960.

Calvino, Italo. *Invisible Cities*. New York: Harcourt, Brace, Jovanovitch, 1978.

Carey, Mathew. *Autobiography*. Brooklyn, NY: Eugene L. Schwaab, 1942.

———. *Essays on the Public Charities of Philadelphia: intended to vindicate the Benevolent Societies of this City from the Charge of encouraging Idleness, and to place in strong Relief, before an enlightened Public, the Sufferings and Oppression under which the greater part of the Females labour, who depend on their industry for a support for themselves and Children*. Philadelphia: J. Clarke, 1829.

———. *Exhibit of the shocking oppression and injustice suffered for sixteen months by John Randel, Jun. Esq., contractor for the eastern section of the Chesapeake and Delaware Canal, from Judge Wright, engineer in chief, and the majority of the board of directors*. Philadelphia: s.n., 1825.

Chura, Patrick. *Thoreau the Land Surveyor.* Gainesville: University Press of Florida, 2010.

Citizens of Philadelphia. *The Claims of the Delaware and Raritan Canal Company, to a Repeal of the Law of Pennsylvania, Passed April 6th, 1825, Entitled, "An Act Relative to the Delaware and Raritan Canal, to be Constructed in New Jersey," And to the Assent of Pennsylvania, Under Just and Equitable Provisions, to the Use of Water of the Delaware*. Philadelphia: Skerrett, 1826.

Clapham, Anthony. "A Short History of the Surveyor's Profession." *Transactions of the Royal Institution of Chartered Surveyors* (December 1949).

Clark, Frank Emerson. *A Treatise on the Law of Surveying and Boundaries*. Indianapolis: Bobbs-Merrill, 1922.

Cmaylo, Dorothy M., Thomas A. Beaver, Kenneth A Regner, and Sheila B. Hoffman, *Images of America: Verona*. Mount Pleasant, SC: Arcadia, 2010.

Cogbill, Charles V., John Burk, and G. Motzkin. "The Forests of Presettlement New England, USA: Spatial and Compositional Patterns Based on Town Proprietor Surveys," *Journal of Biogeography* 29, no. 10–11 (October 2002): 1279–1304.

Cogliano, Francis D. *American Maritime Prisoners in the Revolutionary War: The Captivity of William Russell*. Annapolis: Naval Institute Press, 2001.

Cohen, Patricia Cline. *A Calculating People: The Spread of Numeracy in Early America.* Chicago: University of Chicago Press, 1982.

Cohen, Paul E. "Civic Folly: The Man Who Measured Manhattan." *Bookman's Weekly,* June 13, 1988, pp. 2511–15.

Comegys, Joseph. *Memoir of John M. Clayton*. Wilmington: Historical Society of Delaware, 1882.

Condict, Jemima. *Her Book, Being a Transcript of the Diary of an Essex County Maid During the Revolutionary War*. Newark, NJ: Carteret Book Club, 1930.

Condit, Carl W. *The Port of New York: A History of the Rail and Terminal System from the Beginnings to Pennsylvania Station.* Chicago: University of Chicago Press, 1980.

Condon, George E. *Stars in the Water: The Story of the Erie Canal*. New York: Doubleday, 1974.

Cornog, Evan. "American Antiquity: How DeWitt Clinton Invented Our Past." *American Scholar* 67, no. 4 (Autumn 1998): 53–61.

———. *The Birth of Empire: DeWitt Clinton and the American Experience, 1769–1828.* New York: Oxford University Press, 1998.

Craven, Wayne. "Asher B. Durand's Career as an Engraver." *American Art Journal* 3, no. 1 (Spring 1971): 39–57.

Cronon, William. *Changes in the Land: Indians, Colonists, and the Ecology of New England*. New York: Hill and Wang, 2003.

———. *Nature's Metropolis: Chicago and the Great West*. New York: Norton, 1991.

———. *Uncommon Ground: Rethinking the Human Place in Nature*. New York: Norton, 1996.

Cudahy, Brian. *Under the Sidewalks of New York: The Story of the Greatest Subway System in the World*. 2nd revised edition. New York: Fordham University Press, 1995.

Curl, James Stevens. *The Victorian Celebration of Death*. Stroud, U.K.: Sutton, 2000.

Danson, Edwin. *Drawing the Line: How Mason and Dixon Surveyed the Most Famous Border in America*. New York: Wiley, 2001.

Daston, Lorraine, and Peter Galison. *Objectivity*. Cambridge, MA: Zone, 2007.

Deák, Gloria Gilda. *Picturing America 1497–1899: Prints, Maps, and Drawings Bearing on the New World Discoveries and on the Development of the Territory That Is Now the United States*. Princeton: Princeton University Press, 1988.

DeWitt, Simeon. *The Elements of Perspective*. Albany: H. C. Southwick, 1813.

———. "Observations on the Eclipse of 16 June, 1806, Made by Simeon De Witt Esq. of Albany, State of New-York, Addressed to Benjamin Rush M. D. to Be by Him Communicated to the American Philosophical Society." *Transactions of the American Philosophical Society* 6 (1809): 300–302.

Dixon, Jefferson Max. "The Central Railroad of Georgia, 1833–1892." Doctoral dissertation, George Peabody College for Teachers, 1953.

Duerr, Hans Peter. *Dreamtime: Concerning the Boundary Between Wilderness and Civilization*. Oxford: Blackwell, 1985.

Ehrenberg, Ralph E. *Pattern and Process: Research in Historical Geography*. Washington, DC: Howard University Press, 1975.

Enge, Per. "Retooling the Global Positioning System." *Scientific American*, May 2004, 90–97.

Ernenwein, Raymond P. *Verona Town History*. New York: Heritage, 1970.

Ernst, Joseph W. "With Compass and Chain: Federal Land Surveyors in the Old Northwest: 1785–1816." Doctoral dissertation, Columbia University, 1958.

Evans, Harold. *They Made America: From the Steam Engine to the Search Engine, Two Centuries of Innovators*. New York: Little, Brown, 2004.

Evans, Nancy Goyne. "The Sans Souci, a Fashionable Resort Hotel in Ballston Spa." *Winterthur Portfolio* 6 (1970): 111–26.

Evans, Paul D. "The Frontier Pushed Westward." In *History of the State of New York*, edited by Alexander C. Flick. New York: New York State Historical Association/Columbia University Press, 1934.

Fortune, Brandon Brame, with Deborah J. Warner. *Franklin & His Friends: Portraying the Man of Science in Eighteenth-Century America*. Washington, DC: National Portrait Gallery/ University of Pennsylvania Press, 1997.

Fowler, John. *Journal of a Tour in the State of New York in the Year 1830; with Remarks on agriculture in those parts most eligible for settlers*. London: Whittaker, Treacher, and Arnot,1831.

Frug, Gerald E. "Property and Power: On the Legal History of New York City." *American Bar Foundation Research Journal* 9, no. 3 (Summer 1984): 673–91.

Gallatin, Albert. *Report of the Secretary of the Treasury, on the subject of public roads and canals: made in pursuance of a resolution of Senate, of March 2, 1807*. Washington, DC: R. C. Weightman, 1808.

Gandy, Matthew. *Concrete and Clay: Reworking Nature in New York City*. Cambridge, MA: MIT Press, 2002.

Gardner, Jean, and Joel Greenberg. *Urban Wilderness: Nature in New York*. New York: Earth Environmental Group, 1988.

Gibbs, W. Wayt. "Ultimate Clocks." *Scientific American*, February 2006, 56–63.

Gifford, George E., Jr., ed. *Cecil County, Maryland, 1608–1850 as seen by some visitors and several essays on local history*. Rising Sun, MD: George E. Gifford Memorial Committee/Calvert School, 1974.

Gilreath, James. "Mason Weems, Mathew Carey and the Southern Booktrade, 1794–1810." *Publishing History* 10 (1981): 27–49.

Gittelman, Philip. *At the Met: Olmsted and Central Park*. Metropolitan Museum of Art, 2000. ABC Video Enterprises, 1983.

Goldman, Joanne Abel. *Building New York's Sewers: Developing Mechanisms of Urban Management*. West Lafayette, IN: Purdue University Press, 1997.

Goodwin, Maud Wilder, Alice Carrington Royce, and Ruth Putnam, eds. *Historic New York*. Series I, Volume I. Port Washington: Ira J. Friedman, 1897.

Gopnik, Adam. "Olmsted's Trip: How Did a News Reporter Come to Create Central Park?" *The New Yorker,* March 31, 1997, pp. 96–104.

Gottlieb, Robert. *Forcing the Spring: The Transformation of the American Environmental Movement*. Washington, DC: Island, 1995.

Gould, Peter, and Rodney White. *Mental Maps*. 2nd edition. Boston: Allen & Unwyn, 1986.

Gray, Ralph D. *The National Waterway: A History of the Chesapeake and Delaware Canal, 1769–1985*. 2nd edition. Urbana: University of Illinois Press, 1989.

Greeley, Horace. *Art and Industry: as represented in the Exhibition at the Crystal Palace New York–1853–4: showing the progress and state of the various useful and esthetic pursuits from the New York Tribune*. New York: Redfield, 1853.

Green, Howard. "Factories in a Ditch: The Men Who Built the Delaware and Raritan Canal." Manuscript. Rutgers University Archives, 1999.

Gronim, Sara S. *Everyday Nature: Knowledge of the Natural World in Colonial New York*. New Brunswick: Rutgers University Press, 2007.

———. "Geography and Persuasion: Maps in British Colonial New York." *William and Mary Quarterly,* 3rd series 58, no. 2 (April 2001): 373–402.

Grutzner, Charles. "City Street Plan Is 150 Years Old: First Map, Filed by 3-Man Group, Laid Out Gridiron Pattern for Manhattan." *New York Times*, March 20, 1961.

Guernsey, R. S. *New York City and Vicinity during the War of 1812–15.* New York: Charles L. Woodward, 1889.

Hackett, David G. *The Rude Hand of Innovation: Religion and Social Order in Albany, New York, 1652–1836*. New York: Oxford University Press, 1991.

Hall, Basil. *Travels in North America in the Years 1827 and 1828*. Edinburgh: Cadell, 1829.

Hall, Marcus. *Earth Repair: A Transatlantic History of Environmental Restoration*. Charlottesville: University of Virginia Press, 2005.

Hall, Stephen S. *Mapping the Next Millennium: The Discovery of New Geographies*. New York: Random House, 1992.

Harley, J. B. "Deconstructing the Map." *Cartographica* 26, no. 2 (Summer 1989): 1–20.

Harrington, Samuel M. *Reports of Cases Argued and Adjudged in the Superior Court and Court of Errors and Appeals of the State of Delaware from the Organization of those Courts under the Amended Constitution; with Reference to Some of the Earlier Cases Vol. 1.* Dover, DE: A. M. Schee, 1837.

Harrison, Robert W. *History of the Commercial Waterways & Ports of the United States: From Settlement to Completion of the Erie Canal*. Davis, CA: U.S. Army Engineer Water Resources Support Center, Institute for Water Resources, 1979.

Hartog, Hendrik. *Public Property and Private Power: The Corporation of the City of New York in American Law, 1730–1870*. Chapel Hill: University of North Carolina Press. 1983.

Harvey, David. "Between Space and Time: Reflections on the Geographical Imagination." *Annals of the Association of American Geographers* 80, no. 3 (September 1990): 418–34.

Harwood, Jeremy. *To the Ends of the Earth: 100 Maps That Changed the World*. Cincinnati: F & W, 2006.

Haskell, Daniel C., ed. *Manhattan Maps: A Co-operative List*. New York: New York Public Library, 1931.

Hatton, C. R. "Randel Maps." Manuscript. Manhattan borough president's office.

Hayman, John C. *Rails Along the Chesapeake: A History of Railroading on the Delmarva Peninsula, 1827–1978*. Salisbury, MD: Marvadel, 1979.

Hecht, Roger W. *The Erie Canal Reader: 1790–1950*. Syracuse: Syracuse University Press, 2003.

Heckscher, Morrison H. *Creating Central Park*. New York: Metropolitan Museum of Art and Yale University Press, 2008.

Heidt, William, Jr.. *Simeon DeWitt: Founder of Ithaca*. Ithaca, NY: DeWitt Historical Society of Tompkins County, 1968.

Herring, Thomas A. "The Global Positioning System." *Scientific American*, February 1996, 44–50.

Hessler, John. "From Ortelius to Champlain: The Lost Maps of Henry David Thoreau." *Concord Saunterer: A Journal of Thoreau Studies,* n.s. 18/19 (October/November 2010): 1–25.

Higgins, Ruth. *Expansion in New York, with Especial Reference to the Eighteenth Century*. Columbus: Ohio State University Press, 1931.

Higgs, Eric. *Nature by Design: People, Natural Process, and Ecological Restoration*. Cambridge, MA: MIT Press, 2003.

Hijiya, James A. "Making a Railroad: The Political Economy of the Ithaca and Owego, 1828–1842." *New York History* 54, no. 2 (April 1973): 145–73.

Hill, George Everett, and George E. Waring, Jr. "Old Wells and Water-Courses of the Island of Manhattan, Part II." In *Historic New York: the first series of the Half Moon Papers*, edited by Maud Wilder Goodwin, Alica Carrington Royce, and Ruth Putnam. New York: Knickerbocker/Putnam's, 1897.

Hindle, Brooke. *Emulation and Invention*. New York: Norton, 1983.

Hirschfeld, Charles. "America on Exhibition: The New York Crystal Palace." *American Quarterly* 9, no. 2 (Summer 1957): 101–16.

Hoare, Michael Rand. *The Quest for the True Figure of the Earth: Ideas and Expeditions in Four Centuries of Geodesy*. Aldershot, U.K.: Ashgate, 2005.

Hodges, Graham Russell. *New York City Cartmen, 1667–1850*. New York: New York University Press. 1986.

Holbrook, Stewart H. *The Old Post Road: The Story of the Boston Post Road*. New York: McGraw-Hill, 1962.

Holden, Edward Singleton. *Sir William Herschel: His Life and Works*. London: W. H. Allen, 1881.

Holloway, Marguerite. "I'll Take Mannahatta." *New York Times*, May 16, 2004.

———. "In Amongst the Green Blades." *The Lion and the Unicorn* 35, no. 2 (April 2011): 132–45.

———. "Seeing the Earth for Its Faults: Geological Tours and Guides Expose the Secrets of New York City and Beyond." *Scientific American*, September 2001.

Holmes, William F. "The New Castle and Frenchtown Turnpike and Railroad

Company, 1809–1830, Part II: Canal Versus Railroad." *Delaware History* 10, no. 2 (October 1962), 152–80.

———. "The New Castle and Frenchtown Turnpike and Railroad Company, 1809–1830, Part III: From Horses to Locomotives." *Delaware History* 10, no. 3 (April 1963), 235–70.

Holt, Glen E. "The Changing Perception of Urban Pathology: An Essay on the Development of Mass Transit in the United States." In *Cities in American History*, edited by Kenneth T. Jackson and Stanley K. Schultz. New York: Knopf, 1972.

Homberger, Eric. *The Historical Atlas of New York City: A Visual Celebration of 400 Years of New York City's History*. New York: Holt, 2005.

Hood, Clifton. *722 Miles: The Building of the Subways and How They Transformed New York*. Baltimore: Johns Hopkins University Press, 1993.

Howard, Benjamin C. *Reports of Cases Argued and Adjudged in the Supreme Court of the United States. January Term, 1844*. Philadelphia: Johnson, 1845.

Howell, George Rogers, *Bi-centennial History of Albany. History of the County of Albany, N. Y. from 1609 to 1886*. New York: W. W. Munsell, 1886.

Hoy, Thorkild. "Thoreau as a Surveyor." In *Plotters and Patterns of American Land Surveying*, edited by Roy Minnick. Rancho Cordova, CA: Landmark Enterprises, 1985.

Hoyt, James. *"The Mountain Society": A History of the First Presbyterian Church, Orange, N.J.* New York: Saxton, Barker, 1860.

Hubbard, Bill, Jr.. *American Boundaries: The Nation, the States, the Rectangular Survey*. Chicago: University of Chicago Press. 2009.

Huler, Scott. *On the Grid: A Plot of Land, an Average Neighborhood, and the Systems That Make Our World Work*. New York: Rodale, 2010.

Isenberg, Andrew C., ed. *The Nature of Cities: Culture, Landscape, and Urban Space.* Rochester: University of Rochester Press, 2006.

Iwamoto, Jirô. "Jori System-Division of Cultivated Land in Ancient Japan." *Dialogues d'Histoire Ancienne* 12, no. 12 (1986): 471–78.

Jackson, Kenneth T. *Crabgrass Frontier: The Suburbanization of the United States*. New York: Oxford University Press, 1985.

———, ed. *The Encyclopedia of New York City*. New Haven: Yale University Press, 1995.

Jackson, Kenneth T., and David S. Dunbar, eds. *Empire City: New York Through the Centuries*. New York: Columbia University Press, 2002.

Jacobs, Jane. *The Death and Life of Great American Cities*. New York: Modern Library, 1993.

James, Preston E., and Geoffrey J. Martin. *All Possible Worlds: A History of Cartographical Ideas*. New York: Wiley, 1972.

Janvier, Thomas A. *In Old New York*. New York: Harper & Brothers, 1894.

Jaye, Michael C., and Ann Chalmers Watts, eds. *Literature and the American Urban Experience: Essays on the City and Literature.* New Brunswick: Rutgers University Press, 1981.

Jenkins, Stephen. *The Old Boston Post Road*. New York: Putnam's, 1913.

Johnson, Hildegard Binder. *Order Upon the Land: The U.S. Rectangular Land Survey and the Upper Mississippi Country*. New York: Oxford University Press, 1976.

———. "Rational and Ecological Aspects of the Quarter Section: An Example from Minnesota." *Geographical Review* 47, no. 3 (July 1957): 330–48.

Johnson, Steven. *The Ghost Map: The Story of London's Most Terrifying Epidemic—and How It Changed Science, Cities, and the Modern World*. New York: Riverhead, 2007.

Johnson, William. *Reports of Cases Argued and Determined in the Supreme Court of Judicature, and in the Court for the Trial of Impeachments and The Corrections of Error, in the State of New York. Vol XIV. Second Edition with additional Notes and References.* New York: Banks & Brothers, 1864.

Johnston, George. *History of Cecil County, Maryland, and the Early Settlements around the Head of Chesapeake Bay and on the Delaware River, with Sketches of Some of the Old Families of Cecil County.* Elkton, MD: George Johnston, 1881.

Jones, Ignatius. *Recollections of Albany and of Hudson with Anecdotes and Sketches of Men and Things*. Albany: Charles Van Benthuysen, 1850.

Judd, Richard W. *The Untilled Garden: Natural History and the Spirit of Conservation in America, 1740–1840*. New York: Cambridge University Press, 2009.

Karas, Nick. *Brook Trout: A Thorough Look at North America's Great Native Trout—Its History, Biology, and Angling Possibilities.* Guilford, CT: Lyons, 2002.

Kassulke, Natasha. "See Wisconsin Through the Eyes of 19th-Century Surveyors: Historic Notebooks Play a Critical Role in the Future of Sustainable Ecosystems." *Wisconsin Natural Resources*, supplement. August 2009.

Kennedy, J. Gerald. *The Astonished Traveler: William Darby, Frontier Geographer and Man of Letters*. Baton Rouge: Louisiana State University Press, 1981.

Kidwell, Peggy A. "The Astrolabe for Latitude 41°N of Simeon de Witt: An

Early American Celestial Planisphere." *Imago Mundi: The International Journal for the History of Cartography*. 61, no. 1 (January 2009): 91–96.

Kiely, Edmond Richard. *Surveying Instruments: Their History and Classroom Use*. New York: Teachers College, Columbia University, 1947.

Kieran, John. *A Natural History of New York City*. New York: Fordham University Press, 1982.

Kirschke, James. *Gouverneur Morris: Author, Statesman and Man of the World*. New York: Thomas Dunne, 2005.

Kline, Benjamin. *First Along the River: A Brief History of the U.S. Environmental Movement*. Lanham, MD: Rowman & Littlefield, 2007.

Klinghoffer, Judith A., and Lois Elkis. "The Petticoat Electors: Women's Suffrage in New Jersey, 1776–1807." *Journal of the Early Republic* 12, no. 2 (Summer 1992): 159–93.

Knowles, Anne Kelly, ed. *Past Time, Past Place: GIS for History*. Redlands, CA: ESRI, 2002.

———, ed. *Placing History: How Maps, Spatial Data, and GIS Are Changing Historical Scholarship*. Redlands, CA: ESRI, 2008.

Koeppel, Gerard. *Bond of Union: Building the Erie Canal and the American Empire*. Cambridge, MA: Da Capo, 2009.

Kohler, Robert E. *Landscape and Labscapes: Exploring the Lab-Field Border in Biology*. Chicago: University of Chicago Press, 2002.

Konefsky, Alfred S., and Andrew J. King, eds. *The Papers of Daniel Webster. Legal Papers, Volume 1: The New Hampshire Practice.* Lebanon, NH: University Press of New England, 1982.

Koolhaas, Rem. *Delirious New York: A Retrospective Manifesto for Manhattan*. New York: Monacelli, 1994.

Koop, Frederick W. *Precise Leveling in New York City. City of New York, Board of Estimate and Apportionment. Office of the Chief Engineer. Executed 1909 to 1914*. Ann Arbor: University of Michigan Library, 2006.

Kostof, Spiro. *The City Shaped: Urban Patterns and Meaning Through History*. Boston: Little, Brown, 1991.

Kouwenhoven, John A. *The Beer Can by the Highway: Essays on What's American About America*. Baltimore: Johns Hopkins University Press, 1961.

Lamb, Martha J. *History of the City of New York: Its Origin, Rise and Progress*. New York: A. S. Barnes, 1877.

Lankton, Larry D. "New Castle and Frenchtown Railroad." *Historic American Engineering Record*. Washington, D.C.: National Park Service, 1976.

Larson, John Lauritz. *Internal Improvement: National Public Works and the Promise of Popular Government in the Early United States*. Chapel Hill: University of North Carolina Press, 2001.

Larson, Scott. "Building Like Moses with Jacobs in Mind: Redevelopment Politics in the Bloomberg Administration." Doctoral dissertation, City University of New York, 2010.

Lee, Hardy Campbell, Winton G. Rossiter, and John Marcham. *A History of Railroads in Tompkins County*. 3rd edition. Ithaca: History Center in Tompkins County, 2008.

Lewis, Gene D. *Charles Ellet, Jr.: The Engineer as Individualist: 1810–1862*. Chicago: University of Illinois Press, 1968.

Lewis, G. Malcolm, ed. *Cartographic Encounters: Perspectives on Native American Mapmaking and Map Use*. Chicago: University of Chicago Press, 1998.

Linklater, Andro. *The Fabric of America: How Our Borders and Boundaries Shaped the Country and Forged Our National Identity*. New York: Walker, 2007.

———. *Measuring America: How an Untamed Wilderness Shaped the United States and Fulfilled the Promise of Democracy*. New York: Walker, 2002.

Longley, Paul A., David J. Maguire, Michael F. Goodchild, and David W. Rhind. *Geographical Information Systems and Science*. New York: Wiley, 2010.

Longstreet, Stephen. *City on Two Rivers: Profiles of New York—Yesterday and Today*. New York: Hawthorn, 1975.

Louv, Richard. *Last Child in the Woods: Saving Our Children from Nature Deficit Disorder*. Chapel Hill: Algonquin, 2005.

Love, John. *Geodaesia: Or, The Art of Surveying and Measuring Land Made Easy*. London: G. G. J. and J. Robinson, 1736.

Lowenthal, David, and Martyn J. Bowden, eds. *Geographies of the Mind: Essays in Historical Geosophy in Honor of John Kirkland Wright*. New York: Oxford University Press, 1976.

Macklin, C., J. Wargaski, M. Edwards, and K. Y. Li. "DATAPLAY: Mapping Game Mechanics to Traditional Data Visualization." Proceedings of Digital Games Research Association. London, 2009.

Maier, Pauline, Merritt Roe Smith, Alexander Keyssar, and Daniel Kevles. *Inventing America: A History of the United States*. New York: Norton, 2005.

Mann, Bruce H. *Republic of Debtors: Bankruptcy in the Age of American Independence*. Cambridge, MA: Harvard University Press, 2002.

Mano, Jo Margaret. "History in the Mapping: Simeon DeWitt's Legacy in New York State Cartography." *Proceedings, Association of American Geographers Middle States Division* 22 (1989): 27–33.

———. "Unmapping the Iroquois: New York State Cartography, 1792–1845." In *The Oneida Indian Journey: From New York to Wisconsin, 1784–1860*, edited by Laurence M. Hauptman and L. Gordon McLester. Madison: University of Wisconsin Press, 1999.

Marcuse, Peter. "The Grid as City Plan: New York City and Laissez-Faire Planning in the Nineteenth Century." *Planning Perspectives* 2 (1987): 287–310.

Marino, Cesare, and Karim M. Tiro, eds. *Along the Hudson and Mohawk: The 1790 Journey of Count Paolo Andreani*. Philadelphia: University of Pennsylvania Press, 2006.

Marsh, George Perkins. *Man and Nature*. Seattle: University of Washington Press, 2003 [1864].

Martin, Geoffrey J. "The Emergence and Development of Geographic Thought in New England." *Economic Geography* 74, Supplement 1 (March 1998): 1–13.

Marx, Leo. *The Machine in the Garden: Technology and the Pastoral Ideal in America*. New York: Oxford University Press, 1964.

McKibben, Bill, ed. *American Earth: Environmental Writing Since Thoreau*. New York: Library of America, 2008.

McLean, Albert F., Jr. "Thoreau's True Meridian: Natural Fact and Metaphor." *American Quarterly* 20, no. 3 (Autumn 1968): 567–79.

Merrill, Alvin. "The First Passenger Railway in America." 1908. http://home.roadrunner.com/~nrw/history/railways/merrill1.html.

Merrill, Jason. "History of the Development of the Early Railroad System of Tompkins County." *Ithaca Journal*, 1915.

Meyer, Henry O. A. "Diamond." *AccessScience*. 2012.

Middleton, William D., George M. Smerk, and Roberta L. Diehl, eds. *Encyclopedia of North American Railroads*. Bloomington: Indiana University Press, 2007.

Miller, Angela. *The Empire of the Eye: Landscape Representations and American Cultural Politics, 1825–1875*. Ithaca: Cornell University Press, 1993.

Miller, Benjamin. "Fat of the Land: New York's Waste." *Social Research* 65, no. 1 (Spring 1998), 75–99.

Mix, David E. E. *Catalogue of Maps and Surveys in the Offices of the Secretary of State, State Engineer and Surveyor, and Comptroller.* Albany: Charles Van Benthuysen, 1859.

Monmonier, Mark. *How to Lie with Maps*. Chicago: University of Chicago Press, 1991.

Moore, Clement Clarke. *A Plain Statement, Addressed to the Proprietors of Real Estate in the City and County of New York by A Landholder*. New York: J. Eastburn/Clayton & Kingsland, 1818.

Morgan, Helen M. *A Season in New York, 1801: Letters of Harriet and Maria Trumbull*. Pittsburgh: University of Pittsburgh Press, 1969.

Morris, Anne Cary, ed. *The Diary and Letters of Gouverneur Morris*. Cambridge, MA: Da Capo, 1970.

Morris, Gouverneur, Simeon DeWitt, and John Rutherford. "Remarks of the Commissioners for Laying Out Streets and Roads in the City of New York, under the act of April 3, 1807." www.library.cornell.edu/Reps/DOCS/nyc1811.htm

Munsell, Joel. *The Annals of Albany*. 10 volumes, Albany: J. Munsell, 1850–1859.

———. *Collections on the history of Albany, from its discovery to the present time, with notices of its public institutions, and biographical sketches of citizens deceased.* 4 volumes. Albany: J. Munsell, 1865–1871.

Nash, Roderick Frazier. *Wilderness and the American Mind*. New Haven: Yale University Press, 1967.

New Jersey Legislature. *Report of the Committee, to whom was referred the subject of the Delaware & Raritan Canal, made to the house of assembly on the fifteenth day of January, eighteen hundred and twenty-nine, and ordered to be printed.* Trenton: Joseph Justice, 1829.

———. Delaware and Raritan Canal Commission. *Report of the Commissioners, Appointed by an Act of the Legislature of the State of New-Jersey, For ascertaining the most eligible route for and the probable expense of, a canal to connect the tide waters of the Delaware with those of the Raritan*. New York: Van Winkle, Wiley, 1817.

New York State. Commissioners for Improvement of Internal Navigation. *Report of the Commissioners Appointed by Joint Resolution of the Honorable the Senate and Assembly of the State of New-York of the 13th & 15th March, 1810, to explore the route of an inland navigation from Hudson's River to Lake Ontario and Lake Erie.* Albany: Southwick and Pelsue, 1811.

Nichols, Isaac. *Biographical and Genealogical History of the City of Newark and Essex County, New Jersey.* New York: Lewis, 1898.

O'Carroll, R. E., G. Masterton, et al. "The Neuropsychiatric Sequelae of Mercury Poisoning: The Mad Hatter's Disease Revisited." *British Journal of Psychiatry* 167, no. 1 (August 1995): 95–98.

Oelschlaeger, Max. *The Idea of Wilderness: From Prehistory to the Age of Ecology*. New Haven: Yale University Press, 1991.

Olson, Sherry H. *Baltimore: The Building of An American City*. Baltimore: Johns Hopkins University Press, 1980.

Ōtani, Ryōkichi. *Tadataka Inō: The Japanese Land-Surveyor.* Tokyo: S. Iwanami, 1932.

Pattison, William D. *Beginnings of the American Rectangular Land Survey System, 1784–1800*. Chicago: University of Chicago Press, 1957.

Perry, W. L. *Scenes in a Surveyor's Life; or a Record of Hardships and Dangers encountered, and amusing Scenes which occurred, in the Operations of a Party of Surveyors in South Florida*. Jacksonville: C. Drew's, 1859.

Peterson, Charles. "The Sofa Elevator." *Journal of the Society of Architectural Historians* 10, no. 3 (October 1951).

Petzold, Charles. "How Far from True North Are the Avenues of Manhattan?" www.charlespetzold.com/etc/AvenuesOfManhattan/.

Pickles, John, ed. *Ground Truth: The Social Implications of Geographic Information Systems*. New York: Guilford, 1995.

———. *A History of Spaces: Cartographic Reason, Mapping and the Geo-Coded World*. New York: Routledge, 2004.

Pomerantz, Sidney I. *New York, An American City 1783–1803: A study of urban life*. New York: Columbia University Press, 1938.

Post, Robert C. "Reflections of American Science and Technology at the New York Crystal Palace Exhibition of 1853." *Journal of American Studies* 17, no. 3 (December 1983): 337–56.

Potter, Eliza. *A Hairdresser's Experience in High Life*. Edited by Xiomara Santamarina. Durham: University of North Carolina Press, 2009 [1859].

Pound, Arthur. *The Golden Earth: The Story of Manhattan's Landed Wealth*. London: Macmillan, 1935.

Raisz, Erwin. "Outline of the History of American Cartography." *Isis* 26, no. 2 (March 1937): 373–91.

Randel, John, Jr. "City of New York, north of Canal street, in 1808 to 1821." In *Manual of the Corporation of the City of New York.* Edited by David T. Valentine, and Samuel J. Willis. New York: E. Jones, 1864.

———. *Description of a Direct Route for the Erie Canal, at its eastern termination: with estimates of its expense, and comparative advantages.* Albany: G. J. Loomis, 1822.

———. *The Elevated Railway, and its Appendages, for Broadway, in the City of New York*. New York: J. M. Ellliot, 1848.

———. *Explanatory Remarks and Estimates of the Cost and Income of the*

Elevated Railway, And Its Appendages. New York: George F. Nesbitt, 1848.

———. "Old New York Revived." *Historical Magazine and Notes and Queries Concerning the Antiquities, History and Biography of America.* July 1867.

________. "Report of the Engineer in Chief of the Ithaca and Owego Railroad Company." *American Railroad Journal.* September 14, 1833.

———. "Report on a dry dock design for the Franklin Institute." *Franklin Journal and American Mechanics' Magazine* 3, no. 1 (January 1827): 3–10.

Reeves, William Fullerton. *The First Elevated Railroads in Manhattan and the Bronx of the City of New York*. New York: New-York Historical Society, 1936.

Regis, Pamela. *Describing Early America: Bartram, Jefferson, Crèvecoeur, and the Rhetoric of Natural History*. DeKalb: Northern Illinois University Press, 1992.

Reichel, William C. *A History of the Rise, Progress, and Present Condition of the Bethlehem Female Seminary. With a Catalogue of Its Pupils, 1785–1858.* Philadelphia: Lippincott, 1858.

Reingold, Nathan, ed. *The Papers of Joseph Henry. Volume 1: December 1797– October 1832. The Albany Years*. Washington, D.C.: Smithsonian Institution, 1972.

———, ed. *Science in Nineteenth-Century America: A Documentary History*. Chicago: University of Chicago Press, 1964.

Report of the Commissioners of the Delaware Rail Road with the Charter of the Company. Dover: Samuel Kimmey, 1837.

Report of the President and Directors to the Stockholders of the Ithaca and Owego Rail Road Company. Ithaca: Mack & Andrus, 1833.

Reps, John W. *The Making of Urban America: A History of City Planning in the United States*. Princeton: Princeton University Press, 1965.

Richards, William C. *A Day in The New York Crystal Palace and how to make the most of it: being a popular companion to the "official catalogue," and a guide to all the objects of special interest in the New York exhibition of the industry of all nations*. New York: Putnam, 1853.

Riker, James. *Revised History of Harlem (City of New York).* New York: New Harlem, 1904.

Rink, Evald. *Technical Americana: A Checklist of Technical Publications Printed Before 1831*. Millwood, NY: Kraus International, 1981.

Ristow, Walter W. *American Maps and Mapmakers: Commercial Cartog-*

raphy in the Nineteenth Century. Detroit: Wayne State University Press, 1985.

Roberts, James A. *New York in the Revolution as Colony and State*. Albany: Brandow, 1898.

Rogers, Elizabeth Barlow. "What Is the Romantic Landscape?" *German Historical Institute Bulletin,* Supplement 4 (2007): 11–24.

Rogers, Elizabeth Barlow, Elizabeth S. Eustis, and John Bidwell. *Romantic Gardens: Nature, Art, and Landscape Design*. Boston: Godine, 2010.

Rose-Redwood, Reuben S. "Critical Cartography and Performativity of 3D Mapping." *GIS Development* 11, no. 4 (April 2007), 40–43.

———. "Encountering *Mannahatta*: A Critical Review Forum." *Cartographica* 45, No 4 (2010): 241-272

———. "From Number to Name: Symbolic Capital, Places of Memory, and the Politics of Street Renaming in New York City." *Social & Cultural Geography* 9, no. 4 (June 2008): 432–52.

———. "Genealogies of the Grid: Revisiting Stanislawski's Search for the Origin of the Grid-Pattern Town." *Geographical Review* 98, no. 1 (January 2008): 42–58.

———. "Geographical Field Note: Re-creating the Historical Topography of Manhattan Island." *Geographical Review* 93, no. 1 (January 2003): 124–32.

———. "Governmentality, Geography, and the Geo-Coded World." *Progress in Human Geography* 30, no. 4 (2006): 469–86.

———. "Indexing the Great Ledger of the Community: Urban House Numbering, City Directories, and the Production of Spatial Legibility." *Journal of Historical Geography* 34, no. 2 (2007): 286–310.

———. "Mythologies of the Grid in the Empire City, 1811–2011." *Geographical Review* 101, no. 3 (2011): 396–413.

———. "Rationalizing the Landscape: Superimposing the Grid upon the Island of Manhattan." Master's thesis, Department of Geography, Pennsylvania State University, 2002.

———. "'Sixth Avenue Is Now a Memory': Regimes of Spatial Inscription and the Performative Limits of the Official City-Text." *Political Geography* 27, no. 8 (2008): 875–94.

———. "The Surveyor's Model of the World: The Uses and Abuses of History in Introductory Surveying Textbooks." *Cartographica* 39, no. 4 (Winter 2004): 45–54.

Rose-Redwood, Reuben, and Li Li. "From Island of Hills to Cartesian Flatland? Using GIS to Assess Topographical Change in New York City, 1819–1999." *The Professional Geographer* 63, no. 3 (2011): 392–405.

Rosenwaike, Ira. *Population History of New York City*. Syracuse: Syracuse University Press, 1972.

Rosenzweig, Roy, and Elizabeth Blackmar. *The Park and the People: A History of Central Park*. Ithaca: Cornell University Press, 1992.

Roy, William. "An Account of the Measurement of a Base on Hounslow-Heath." *Philosophical Transactions of the Royal Society of London* 75 (1795): 385–478.

Rubin, Julius. "Canal or Railroad? Imitation and Innovation in the Response to the Erie Canal in Philadelphia, Baltimore, and Boston." *Transactions of the American Philosophical Society,* n.s. 51, no. 7 (1961): 1–106.

Rundell, Maria. *American Domestic Cookery, Formed on Principles of Economy, for the Use of Private Families by an Experienced Housekeeper.* New York: Evert Duyckinck, 1823.

Ryan, Mary P. *Cradle of the Middle Class: The Family in Oneida County, New York, 1790–1865*. Cambridge: Cambridge University Press, 1981.

Sanderson, Eric W. *Mannahatta: A Natural History of New York City*. New York: Abrams, 2009.

Sanderson, Eric W., and Marianne Brown. "Mannahatta: An Ecological First Look at the Manhattan Landscape Prior to Henry Hudson." *Northeastern Naturalist* 14, no. 4 (2007): 545–70.

Sanderson, Eric W., et al. "The Human Footprint and the Last of the Wild." *BioScience* 52, no. 10 (October 2002): 891–904.

Sayre, Robert. *Recovering the Prairie*. Madison: University of Wisconsin Press, 1999.

Schama, Simon. *Landscape and Memory*. New York: Vintage, 1996.

Schecter, Barnet. *The Battle for New York: The City at the Heart of the American Revolution*. New York: Walker, 2002.

Schein, Richard H. "Urban Origin and Form in Central New York." *Geographical Review*. 81, no. 1. (January 1991): 52–69.

Schudson, Michael. *Discovering the News: A Social History of American Newspapers*. New York: Basic, 1978.

Schuyler, David. *The New Urban Landscape: The Redefinition of City Form in Nineteenth-Century America*. Baltimore: Johns Hopkins University Press, 1986.

Scobey, David M. *Empire City: The Making and Meaning of the New York City Landscape*. Philadelphia: Temple University Press, 2003.

Scott, Ira W. *Albany Directory, for the Year 1826: Containing an Alphabetical List of Residents Within the City, and a Variety of Miscellaneous Matter, for the Benefit of the Public*. Albany: Webster & Wood, 1826.

Shabecoff, Philip. *A Fierce Green Fire: The American Environmental Movement*. New York: Hill and Wang, 1993.

Shami, Charles Henry. "Charles Ellet, Jr.: Early American Economic Theorist and Econometrician, 1810–1862: An Analytical Exposition of His Theories." Doctoral dissertation, Columbia University, 1968.

Shanor, Rebecca Read. *The City That Never Was: Two Hundred Years of Fantastic and Fascinating Plans That Might Have Changed the Face of New York City*. New York: Penguin, 1988.

———. "New York's Paper Streets: Proposals to Relieve the 1811 Gridiron Plan." New York Neighborhood Studies, Working Paper #3. New York: Columbia University, 1982.

Sheriff, Carol. *The Artificial River: The Erie Canal and the Paradox of Progress, 1817–1862*. New York: Hill and Wang, 1996.

Short, John Rennie. *Imagined Country: Environment, Culture and Society*. London: Routledge, 1991.

———. *Representing the Republic: Mapping the United States 1600–1900*. London: Reaktion, 2001.

Shorto, Russell. *The Island at the Center of the World: The Epic Story of Dutch Manhattan and the Forgotten Colony That Shaped America*. New York: Vintage, 2005.

Simutis, Leonard J. "Frederick Law Olmsted Sr.: A Reassessment." *AIP Journal* 38, no. 5 (September 1972): 276–84.

Smart, Charles E. *The Makers of Surveying Instruments in America Since 1700*. Troy, NY: Regal Art, 1962–1967.

Smith, James R. *Introduction to Geodesy: The History and Concepts of Modern Geodesy*. New York: Wiley, 1997.

Snyder, John P. *The Mapping of New Jersey: The Men and the Art*. New Brunswick: Rutgers University Press, 1973.

———. *The Mapping of New Jersey in the American Revolution*. Trenton: New Jersey Historical Commission, 1975.

Society for the Promotion of Useful Arts. *Charter and By-Laws of the Society for the Promotion of Useful Arts; Together with a list of the members and the standing committees*. Albany: Websters and Skinners, 1815.

———. *Transactions of the Society for the Promotion of Useful Arts, in the State of New-York*. Albany: John Barber, 1807.

———. *Transactions of the Society for the Promotion of Useful Arts, in the State of New-York, Volume III*. Albany: Websters and Skinners, 1814.

———. *Transactions of the Society for the Promotion of Useful Arts, in the State of New-York, Volume IV–Part I*. Albany: Websters and Skinners, 1816.

———. *Transactions of the Society for the Promotion of Useful Arts, in the State of New-York, Volume IV–Part II*. Albany: Websters and Skinners, 1819.

Sokoloff, Kenneth L., and B. Zorina Kahn. "The Democratization of Invention During Early Industrialization: Evidence from the United States, 1790–1846." *Journal of Economic History* 50, no. 2 (June 1990): 363–78.

Spann, Edward K. "The Greatest Grid: The New York Plan of 1811." In *Two Centuries of American Planning*. Edited by Daniel Schaffer. London: Mansell, 1988.

———. *The New Metropolis: New York City, 1840–1857*. New York: Columbia University Press, 1981.

Spirn, Anne Whiston. "The Authority of Nature: Conflict and Confusion in Landscape Architecture." In *Nature and Ideology: Natural Garden Design in the Twentieth Century*. Edited by Joachim Wolschke-Bulmahn. Washington, D.C.: Dumbarton Oaks, 1997.

———. "Constructing Nature: The Legacy of Frederick Law Olmsted." In *Uncommon Ground: Rethinking the Human Place in Nature*. Edited by William Cronon. New York: Norton, 1996.

Stilgoe, John R. *Common Landscape of America: 1580 to 1845*. New Haven: Yale University Press, 1982.

Stokes, I. N. Phelps. *The Iconography of Manhattan Island, 1498–1909*. Six volumes. New York: R. H. Dodd, 1915–1928.

Stover, John. *American Railroads*. Chicago: University of Chicago Press, 1997.

Stradling, David. *Making Mountains: New York City and the Catskills*. Seattle: University of Washington Press, 2007.

———. *The Nature of New York: An Environmental History of the Empire State*. Ithaca: Cornell University Press, 2010.

Swan, Herbert S. "The Minor Street." *Journal of Land & Public Policy Utility Economics* 9, no. 3 (August 1933): 297–305.

Tanner, H. S. *The American Traveller; or Guide Through the United States. Containing brief notices of the several states, cities, principal towns, canals and rail roads, &c. with Tables of Distances, by stage, canal, and steam boat routes. The whole Alphabetically arranged, with direct Reference to the accompanying Map of the Roads, Canals, and Railways of the United States*. Philadelphia: H. S. Tanner, 1839.

Taylor, George Rogers. *The Transportation Revolution: 1815–1860*. New York: Rinehart & Company, 1951.

Teaford, Jon C. "Caro versus Moses, Round Two: Robert Caro's *The Power Broker*." *Society for the History of Technology* 49, no. 2 (April 2008): 442–48.

Thorn, John, ed. *New York 400: A Visual History of America's Greatest*

City with Images from the Museum of the City of New York. Philadelphia: Running Press, 2009.

Timreck, T. W. *Frederick Law Olmsted and the Public Park in America*. Metropolitan Museum of Art, 2008. Filmstrip.

Tobey, Ronald. *Saving the Prairies: The Life Cycle of the Founding School of American Plant Ecology, 1895–1955.* Berkeley: University of California Press, 1981.

Trachtenberg, Alan. "The Rainbow and the Grid." *American Quarterly* 16, no. 1 (Spring 1964): 3–19.

Transaction of the American Institute of the City of New-York for the Year 1855. Albany: C. Van Benthuysen, 1856.

Valentine, David T., and Samuel J. Willis. *Manual of the Corporation of the City of New York.* New York: E. Jones, 1864.

"Variation of the Magnetic Needle." *American Journal of Science and Arts* 16, no. 1 (July 1829): 60–62.

Viele, Egbert L. *The Topography and Hydrology of New York*. New York: Robert Craighead, 1865.

Von Gerstner, Franz A. R. *Early American Railroads, 681, 724; Reports of the Presidents, Engineers-in-Chief and Superintendents, of the Central Rail-Road and Banking Company of Georgia, No. 1 to 19 Inclusive, with Report of Survey by Alfred Cruger, and the Charter of the Company*. Savannah: John M. Cooper, 1854.

Walker, Herbert T. "Early History of the Delaware, Lackawanna & Western Railroad and Its Locomotives (Ithaca & Owego etc)." *Railroad Gazette* (May 30, 1902): 388–89.

Warner, Deborah Jean. "Optics in Philadelphia During the Nineteenth Century." *Proceedings of the American Philosophical Society* 129, no. 3 (September 1985): 291–99.

———. "Terrestrial Magnetism: For the Glory of God and the Benefit of Mankind." *Osiris* 9 (1994): 67–84.

———. "True North—And Why It Mattered in Eighteenth-Century America." *Proceedings of the American Philosophical Society* 149, no. 3 (September 2005): 372–85.

Weeks, William R. "Interesting Data From the Papers of J. Randel, Jr." *Numismatist*27, no. 11 (December 1915): 443.

Whitehead, John. *The Judicial and Civil History of New Jersey.* Boston: Boston History Company, 1897.

Whitehill, Walter. *Boston: A Topographical History*. Cambridge, MA: Harvard University Press, 2000.

Whitford, Noble E. *History of the Canal System of the State of New York:*

Together with Brief Histories of the Canals of the United States and Canada. Albany: Brandow, 1906.

Whittemore, Henry. *The Founders and Builders of the Oranges*. Newark, NJ: L. J. Hardham, 1896.

Wilson, Donald A. *Plotters and Patterns of American Land Surveying: A Collection of Articles from the Archives of the American Congress on Surveying and Mapping*. Edited by Roy Minnick. Mount Pleasant, SC: Landmark Enterprises, 1986.

Wilson, E. O. *Biophilia: The Human Bond with Other Species*. Cambridge, MA: Harvard University Press, 1984.

Winchester, Simon. *The Map That Changed the World: William Smith and the Birth of Modern Geology.* New York: HarperCollins, 2001.

Wood, Denis. *The Power of Maps*. New York: Guilford, 1992.

Wood, Denis, and John Fels. *The Natures of Maps: Cartographic Constructions of the Natural World*. Chicago: University of Chicago Press, 2008.

Worster, Donald. *Nature's Economy: A History of Ecological Ideas*. Cambridge: Cambridge University Press, 1994.

Worth, Gorham A. *Random Recollections of Albany, from 1800 to 1808: With Some Additional Matter.* Albany: Charles Van Bethuysen, 1850.

Worthen, W. E., ed. *Appleton's Cyclopaedia of Technical Drawing.* New York: D. Appleton, 1896.

Wyld, Lionel D., ed. *The Erie Canal, 150 Years.* Rome, NY: Oneida County Erie Canal Commemoration Commission, 1967.

Young, Terence. "Modern Urban Parks." *Geographical Review* 85, no. 4 (1995): 535–51.

Young, William C. "Surveying the Erie Canal 1817." Address to the Buffalo Historical Society, 1866. Syracuse: Collection of the Erie Canal Museum.

Zagarri, Rosemarie. *Revolutionary Backlash: Women and Politics in the Early American Republic*. Philadelphia: University of Pennsylvania Press, 2007.

Zelenak, Scott S. "The Xs and Ys of the Big Apple: A Report on the Coordinate and Elevation Origins in the Greater New York City Area." *POBonline,* January 29, 2001. www.pobonline.com/Articles/Features/1a1a817cac0f6010VgnVCM100000f932a8c0____.

INDEX

Page numbers in *italics* refer to illustrations.
Page numbers beginning with 307 refer to notes.